Springer Theses

Recognizing Outstanding Ph.D. Research

Aims and Scope

The series "Springer Theses" brings together a selection of the very best Ph.D. theses
from around the world and across the physical sciences. Nominated and endorsed by two recognized specialists, each published volume has been selected for its scientific excellence and the high impact of its contents for the pertinent field of research. For greater accessibility to non-specialists, the published versions include an extended introduction, as well as a foreword by the student's supervisor explaining the special relevance of the work for the field. As a whole, the series will provide a valuable resource both for newcomers to the research fields described, and for other scientists seeking detailed background information on special questions. Finally, it provides an accredited documentation of the valuable contributions made by today's younger generation of scientists.

Theses may be nominated for publication in this series by heads of department at internationally leading universities or institutes and should fulfill all of the following criteria

- They must be written in good English.
- The topic should fall within the confines of Chemistry, Physics, Earth Sciences, Engineering and related interdisciplinary fields such as Materials, Nanoscience, Chemical Engineering, Complex Systems and Biophysics.
- The work reported in the thesis must represent a significant scientific advance.
- If the thesis includes previously published material, permission to reproduce this must be gained from the respective copyright holder (a maximum 30% of the thesis should be a verbatim reproduction from the author's previous publications).
- They must have been examined and passed during the 12 months prior to nomination.
- Each thesis should include a foreword by the supervisor outlining the significance of its content.
- The theses should have a clearly defined structure including an introduction accessible to new PhD students and scientists not expert in the relevant field.

Indexed by zbMATH.

More information about this series at https://link.springer.com/bookseries/8790

Po Hong

Water Content Estimation and Control of PEM Fuel Cell Stack and the Individual Cell in Vehicle

Doctoral thesis accepted by
Tsinghua University, Beijing

清華大学出版社
TSINGHUA UNIVERSITY PRESS

Springer

Author
Dr. Po Hong
State Key Laboratory of Automotive Safety and Energy, School of Vehicle and Mobility
Tsinghua University
Beijing, China

Supervisor
Prof. Minggao Ouyang
State Key Laboratory of Automotive Safety and Energy, School of Vehicle and Mobility
Tsinghua University
Beijing, China

ISSN 2190-5053 ISSN 2190-5061 (electronic)
Springer Theses
ISBN 978-981-16-8813-3 ISBN 978-981-16-8814-0 (eBook)
https://doi.org/10.1007/978-981-16-8814-0

Jointly published with Tsinghua University Press
The print edition is not for sale in China (Mainland). Customers from China (Mainland) please order the print book from: Tsinghua University Press.

This Springer imprint is published by the registered company Springer Nature Singapore Pte Ltd.
The registered company address is: 152 Beach Road, #21-01/04 Gateway East, Singapore 189721, Singapore

Supervisor's Foreword

The hydrogen Proton Exchange Membrane (PEM) fuel cell plays an important role in realizing a sustainable society with by-product water and less emission of carbon dioxide. Therefore, efforts have been made for decades toward demonstration and commercialization of vehicles powered by the system composed of the PEM fuel cell and essential auxiliaries. Research of the PEM fuel cell system is a comprehensive subject of material, electrochemistry, system control, thermal dynamics, fluid dynamics, electrical and electronics, mechanics, and so on. Until now, the knowledge of scientific and technological obstacles for the application of the PEM fuel cell have been adequately identified and shared by scholars and frontline engineers all over the world. Among obstacles, modeling and estimation of water content in a fuel cell is the most challenging and any advancement would be critical and significant.

Looking through representative references in this field, we find that excellent research results by scholars from different academic fields are presented in a dispersed form and it costs effort and time for students to get started and establish their knowledge and experience effectively. Meanwhile, inspired by continuous technology development of the fuel cell by Toyota Motor Corporation, a doctoral thesis for looking into the water content inside based on the previous contribution is put forward by our research group. Dr. Hong shows great interest in this object and during his bachelor's degree study in the department of automotive engineering, his knowledge on multiple disciplines enriched and initially practiced in terms of electrical and electronics, mechanics, system control, and electrochemistry. The interdisciplinary knowledge contributes to his research in the fuel cell.

Throughout the doctoral thesis, Dr. Hong presents a comprehensive model of water content in the cathode catalyst layer of the fuel cell, based on a physical phenomenon that gas tends to be transported in the void space other than in the liquid phase. This phenomenon results from orders of magnitude larger of resistance for gas transport in the liquid phase than that for gas transport in the void space. The derived profile of liquid water content along the simplified pore agrees well with research results by Toyota Motor Corporation presented later. Besides, Dr. Hong presents a parallel topology of DC/DC converters for imposing AC current excitation on the fuel cell, and a comprehensive scheme of analog signal processing circuits, digital signal processing techniques, and error compensation methods for weak-signal extraction and resistance calculation. Then the approach to online AC impedance measurement of any fuel cell in the stack is experimentally validated with high accuracy and comparable time cost for each measurement trigger with Toyota Motor Corporation. The synthesis of frequency characteristics of the electrical circuit and error compensation for different electrical circuits, as well as small signal model of power electronic circuits, are the key to his research.

On the other hand, Dr. Hong understands the connection between water management of the fuel cell and drying out of clothes in various situations. Applying the mechanism of drying out of clothes to water management of the fuel cell is quantified based on water equilibrium and saturation pressure of water vapor at the cathode outlet. Combing estimation of water content, control-oriented model of the fuel cell system, and online AC impedance measurement together, the system-level water management strategy is established and validated with experiment. The synthesis of frequency characteristics of the fuel cell system with respect to dynamics of mechanical rotation, gas transport, electronic-control valve, and water transport contributes to his research.

This doctoral thesis is a good example of multidisciplinary research and comprehensive application. We would like to recommend this book to readers who are interested in fuel cell and those who try to think about nature as a unity in a philosophical way. We would appreciate any suggestion in the book content and hope to keep in contact with readers all over the world for broader knowledge and our better future.

Beijing, China
December 2021

Minggao Ouyang

Abstract

With the energy crisis and environmental pollution growing more serious, saving energy and being environmentally friendly are attracting our attention. Proton Exchange Membrane Fuel Cell (PEMFC), as an electrochemical energy conversion device, is a very promising alternative for improving our life and environment, thanks to its low emission, low operating noise and high efficiency, etc. Government and enterprises all over the world start fundamental research projects on the PEM fuel cell, implement the demonstration of fuel cell vehicles, and even promote commercialization of the fuel cell system with ever greater efforts, while the fuel cell durability and system cost still limit its marketization process.

To improve the durability, the fuel cell must operate in optimized states continuously but the water content inside a fuel cell is full of uncertainty. Managing the water content also becomes a key problem for the design and control of the fuel cell engine. However, the PEM fuel cell is a complex system with gas, water, electricity, heat, and force being coupled. Estimating water content based on the mathematical model of multi-physics lacks effective information to enhance accuracy and to solve the model both analytically or numerically requires much computation efforts; AC impedance provides significant support for water estimation but its implementation is a great challenge. In hence, it is necessary to study how to combine the multi-physics model and AC impedance together to have a better estimation of water content and optimize the system control.

From the aspect of electrical characteristics, this paper proposes a model for water content inside the Membrane Electrode Assembly (MEA) of a fuel cell. The mathematical connection between the polarization curve and equivalent circuit model proves the importance of MEA and its feasibility to study water content. Optimizing the structure of MEA realizes the internal water content recirculation of a fuel cell and improves its performance under middle or lower current density. This model quantifies the influence of water content on the performance of MEA and variation of equivalent circuit model is an excellent indicator of water content.

Besides, this paper put forwards the comprehensive online AC impedance measurement method, including current excitation method, weak voltage and current signal processing method, and method for analyzing measurement error. Experiment validates the accuracy of the proposal and it can be applied to the fuel cell power-train system. This paper analyzes the sensitivity of the water content of a fuel cell with self-humidifying MEA to operating conditions experimentally and the high frequency impedance and statistical characteristics are proposed as indicators of water content. This paper simulates the online water content estimation algorithm for a fuel cell with MEA of no self-humidifying ability.

In addition, this paper establishes the dynamic model of the air supply system of a fuel cell engine and decouples the closed-loop control of the air supply system and the water content estimation. The experiment on a fuel cell system validates the proposed method for searching optimized operating conditions and the water management strategy.

The innovation of this paper includes the following aspects: A water content model of the MEA is proposed which quantifies the influence of water content on performance and electrical characteristic of a fuel cell; The online AC impedance measurement method for a fuel cell stack and each cell is proposed and experimentally validated with comparable accuracy to the commercial AC impedance equipment and application feasibility; The water management strategy oriented for the fuel cell stack and each cell is put forward and experiment validates the closed-loop control of water content in the MEA and improvement of performance and uniformity of the fuel cell. The work in this paper explores the combination of the multi-physics model and AC impedance to estimate water content online and optimize the system control, which may have certain meaning for application and research of a fuel cell system.

Keywords PEMFC · Membrane electrode assembly · AC impedance · Water content estimation · Water management strategy

Acknowledgements

Here, I want to express my sincere thanks to the teachers (Profs. Jianqiu Li and Liangfei Xu) that help me finish the research thesis for the degree of Doctor of Philosophy. After the five-year study and life full of pain and happiness, I find myself the lucky boy to experience so much that is beyond my imagination. Also, I want to express my sincere thanks to my parents and wife who support me unshakably and with great care.

Fortunately, I have so many lovely schoolmates in our research laboratory, their optimistic attitude to life moves me, and their research contributes to enriching my knowledge. Every day, we help each other and enjoy life together and that's an unforgettable memory.

I also want to express my sincere thanks to colleagues from companies and research institutes for helping provide system test equipment. Through discussion with colleagues in our field, my knowledge is updated day by day and practice makes perfect.

Contents

Chapter 1
Introduction

1.1 Introduction to Research Background and Research Thesis

1.1.1 Research Background

Ever since the twenty-first century, the world population experiences significant increase and the growing consumer demand for better life leads to burst consumption of nature resources. However, the formation of nonrenewable energy like coal and petroleum needs thousands of years and their available storage is challenged with visible great reduction every year for the last two decades. It is predictable that current available energy would not satisfy people's requirement sooner or later. Report by the world energy agency [1] shows that the average new exploration of coal every year in the following 24 years would be decreased to 25% of that in the past 7 years while the exploration of gas energy will be kept stable due to increased found natural gas and combustible ice and so on (Fig. 1.1). On the contrary, the total amount of nuclear energy rises slowly and the contribution of average increased renewable energy every year would be larger than that of all other energy resources together. The renewable energy has been increased by 30% more than that in the past 7 years.

Meanwhile, the great energy consumption results in the emerging pollution like greenhouse gases, nitrogen oxides and sulfide gases. To save the daily-severe environment and maintain sustainable development of energy, governments all over the world start paying attention to renewable energy such as solar energy, wind energy and biomass energy. The research and application of renewable energy shows prosperity with lots of investment. Among countries, China plays an important role and the government takes measures to promote the development of renewable energy in many fields. The architecture of national energy demand [1] is optimized step-by-step and the dependence on single energy resource is reduced for energy safety (Fig. 1.2). The energy saving and the environment protection (Fig. 1.3) are guaranteed at the same time [1].

P. Hong, *Water Content Estimation and Control of PEM Fuel Cell Stack and the Individual Cell in Vehicle*, Springer Theses, https://doi.org/10.1007/978-981-16-8814-0_1

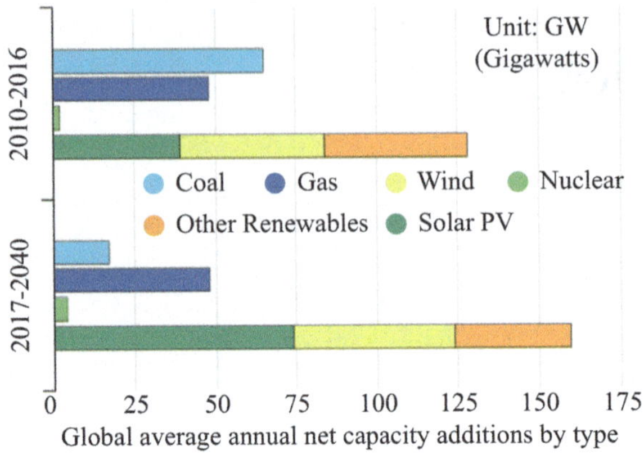

Fig. 1.1 Global average annual net capacity additions by type

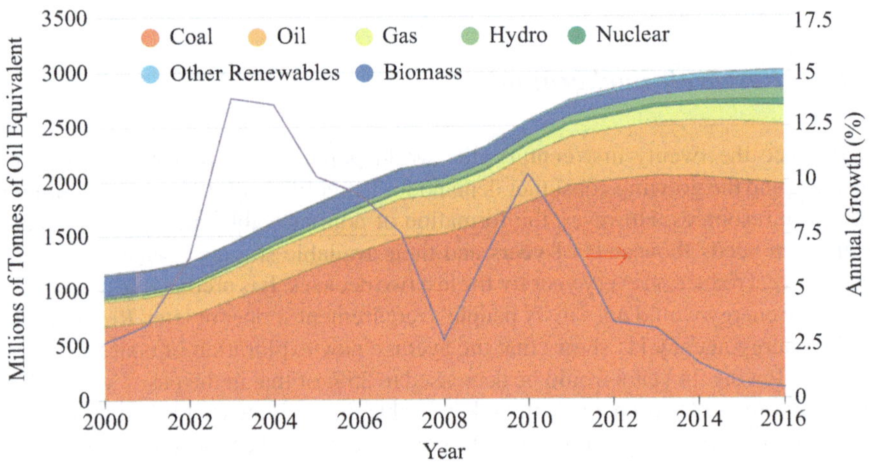

Fig. 1.2 Energy demand and annual growth

Among various renewable energy, the hydrogen energy is the most special. To be specific, the energy density of hydrogen is the highest among gases and the by-product is just water to be environmentally friendly. The source of hydrogen is rich and with the development of low temperature solid-state proton exchange membrane, the application, demonstration and promotion of hydrogen fuel cell in transportation is very promising. The hydrogen proton exchange membrane fuel cell is an electrochemical reaction device with two separated reaction electrodes. It converts the chemical energy to the electrical energy to electrical load and even though the efficiency loss exists for the device, the synthesis efficiency is comparable to that of the petroleum. Besides, the operated device is quiet and low emissions.

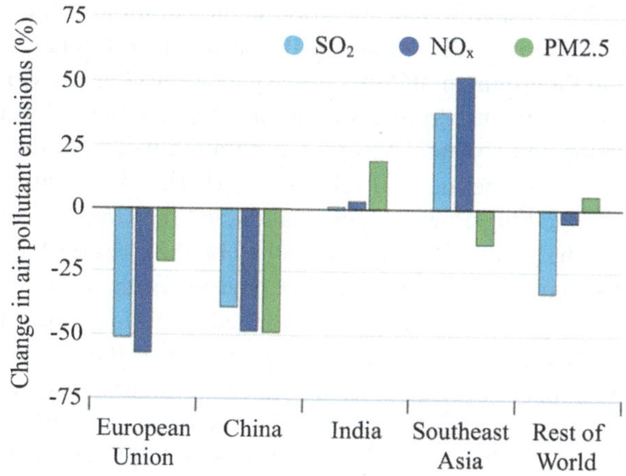

Fig. 1.3 Change in air pollutant emissions

Creating the hydrogen-energy-based society and quickening commercialization of the fuel cell vehicle are an important target for most countries nowadays. The policy guidance is being improved to support development of the complete product chain. Manufacturers, research centers and universities are devoted to advancement of the hydrogen fuel cell related technology. Great progress has been made for the last two decades.

The Ballard Power Corporation [2] developed a fuel cell stack using the compressor to supply air in 1986 for the first time and a 5 kW fuel cell system was their great success in this field. In 1991, the Ballard Power Corporation cooperated with General Motors Corporation and the Department of Energy of U.S.A. to develop the fuel cell system for passenger vehicle. Then they designed a 90 kW fuel cell system for the public transit bus and the demonstration operation was launched in Vancouver. The strategy cooperation contract was signed between the Ballard Power Corporation and the Daimler-Benz Corporation to improve the specific power density of the fuel cell stack and two years later the index reached 700 W/kg.

During the first decade of the twenty-first century, the operation demonstration was launched all over the world, 30 public transit buses were deployed in Europe cities, 3 public transit buses deployed in Australia and 3 public transit buses deployed in China. For special situations, the power of the fuel cell stack with water cooling ranges from 4 to 21 kW and that with air cooling ranges from 400 W to 3.3 kW, which facilitates companies for fuel cell system integration and the OEM for system application. For stable power station, the available power of the fuel cell stack is larger than 30 kW. For powertrain system of vehicles, the available power of the fuel cell stack ranges from 30 to 120 kW. The membrane electrode assembly (MEA) is the key component in a fuel cell and the Ballard Power Corporation invented the advanced MEA with complete manufacturing method, the key components for the fuel cell system and the system integration techniques, which contributes to durability for as long as 10,000 h. Until 2017, the total operation mileage of public transit buses by the Ballard is far more than 10 million km.

The research and operation demonstration in the U.S.A. is more attractive and fruitful. With the support of Department of Energy, Department of Transportation and Environmental Protection Agency, the General Motors Corporation, Ford Corporation, Toyota Motor Corporation, Daimler-Benz Corporation, Nissan Motor Corporation and Hyundai Motors Corporation were all devoted to the technology demonstration of the fuel cell vehicles [3, 4]. They have cultivated many famous companies like the Plug-Power Incorporation and the United Technologies Incorporation.

The project Driveway launched by the General Motors Corporation released over 100 fuel cell vehicles with the Chevrolet Brand Equinox to customers. Until 2009, the total operation mileage has reached 1 million miles. Later on, the General Motors Corporation developed next generation of the fuel cell system and in comparison with that of the Equinox, the system volume was decreased by 50%, the system mass decreased by 220 pounds and most importantly the usage of platinum is reduced by almost 70%. According to the price of platinum in the global market, cost of the stack is saved for more than 10 thousand RMB for a 100 kW fuel cell system [5]. The system cost is also significantly reduced. In 2006, the U.S.A. launched the project for the public transit buses and durability of the fuel cell vehicle is as long as 11,000 h based on the real-time bus data [6].

The fuel cell technology by manufacturers in Japan and Korea is top of the world and the representatives are the Toyota Motor Corporation, Nissan Motor Corporation and Hyundai Motor Corporation. In 2008, the Toyota Motor Corporation released the FCHV-adv passenger car and the fuel cell system can be started with environment temperature at −37 °C. The car can run for 830 km with once refueling and the hydrogen consumption per hundred kilo meters is just 0.7 kg [7].

To promote spread of the technology, the Toyota Motor Corporation announced to share more than 5,000 patents of fuel cell system and its application. In 2014, they released a grand new fuel cell vehicle named MIRAI and the specific power density is 3.1 kW/L. Besides, the hydrogen refueling can be finished within 5 min and the nominal pressure of the hydrogen tank is 70 MPa which significantly improve internal space for passengers. The declared durability is for 5,000 h. The bipolar plate is designed to be 3-D mesh shape, the humidifier is cancelled and key components of the air and hydrogen supply subsystem are integrated with pack of the fuel cell stack. The hydrogen recirculation pump is capable of breaking ice during the cold startup and the DC/DC converter together with the cell voltage monitor can measure the AC impedance of the fuel cell to control water content. The fuel cell system is also applied to the public transit bus and it proves the flexibility of the fuel cell system.

The fuel cell technology by the Nissan Motor Corporation is also very advanced. To be specific, the developed 90 kW fuel cell stack in 2011 weighs only 43 kg and the specific power density is enhanced to be 2.5 kW/L in 2012, which is just a little lower than that by the Toyota Motor Corporation [8]. The fuel cell vehicle named FCX Clarity by Honda Motor Corporation can realize the startup with environment temperature below −30 °C and the nominal driving range is 630 km. Besides, the National Department of Transportation formulates policies to establish the foundation of hydrogen refueling station all over the country and this is the necessity for promotion of fuel cell vehicles to end customers.

Hyundai Motor Corporation in Korea has started research of the hydrogen fuel cell since 2002. They manufactured 32 fuel cell vehicles by assembling fuel cell stack provided by Ballard Power Corporation and the vehicle is a SUV to accommodate the hydrogen tank in 2005. One year later, the corporation released their first generation of the fuel cell stack on their own and manufactured 30 SUVs and 4 buses equipped with the self-developed stack. From the year 2009 to 2012, the self-made fuel cell stack is improved and the second generation of the stack is released. In 2012, the third generation of the fuel cell driven SUV and bus is released and the global promotion and demonstration starts. In 2013, they announce to set up production line for one thousand SUVs per year with almost two year earlier and the name of the SUV is TUCON ix35. The SUV is equipped with a 100 kW fuel cell stack, a 24 kW lithium-ion battery pack and the 70 MPa hydrogen tank for 5.6 kg hydrogen. The NEDC circulation test shows a mileage for 588 km [9].

The Europe Union has made great progress since the 6th and 7th framework project for the hydrogen fuel cell, especially in the fuel cell stack and the hydrogen energy application. With support of the CUTE project and other projects launched by the Europe Union, many cities participate in operation demonstration of public transit buses. In 2013, the CHIP project is started and the durability and reliability of the fuel cell system is enhanced. Besides, the system cost is reduced. Until 2015, the national network of the hydrogen refueling station has been founded in Germany, which contributes to promotion of the fuel cell vehicles. The Daimler Group Corporation takes part in the HyFLEET-CUTE project for clean-energy transportation from the year 2003 to 2009 and 36 fuel cell city buses named Citaro and manufactured by Mercedes-Benz are put into operation. The total operation time has reached 140 thousand hours and the total mileage reached 2.2 million miles. Since 2009, they released the second generation of the fuel cell city bus and most indexes of the fuel cell stack has been top of the world. Along with the enhanced performance, the system cost is greatly reduced and the durability increased for as long as 12,000 h [10]. In 2011, the Daimler-Benz Corporation starts their global show of the fuel cell vehicle and the commercialization of the fuel cell vehicle is validated with comparable performance with that of the internal combustion engine.

China Government launches some major national projects to support fundamental research and application research for development of new energy, including the electric vehicle project of the tenth five-year plan, the energy saving and new energy vehicle project of the eleventh five-year plan and the development of electric vehicle project of the twelfth five-year plan. All these national projects originate from the famous 863 plan. With support of these projects, the technology of the fuel cell vehicle has been advanced significantly. Many corporations overcome lots of technology problems and own the core technology related to critical parts of the fuel cell system, the powertrain system driven by the fuel cell system and the manufacture of fuel cell vehicle. The fuel cell city bus serves players and organizations during the 2008 Beijing Olympic Games and during the 2010 Shanghai World Expo with good operation demonstration. The UNDP project and the ten-city-one-thousand-vehicle project for fuel cell city bus make a great success and people in China begins to pay much attention to the application of fuel cell vehicles.

Table 1.1 Parameters of typical fuel cell engines all over the world

Index	GM	Toyota	Honda	Benz	Ballard	Hydrogenics	US-Hybrid
SPD (kW/L)	1.5	3.1	2.1	–	0.17	0.20	0.20
NP (kW)	92	114	≥ 100	100	85	85	130
Life (h)	5,500	5,000	5,000	5,000	20,000	20,000	20,000
Pt load (g)	30	20	12	20	–	–	–
Stack mass (kg)	130	56	–	–	244	275	474
CS/CSU capability (°C)	−40	−30	−30	−25	−40 CS/−20 CSU	−40 CS/ 20CSU	−40 CS
BP material	Metal	Metal	Metal	Metal	Carbon		Carbon
Operating pressure	High	High	High	High	High		Low

During the eleventh five-year plan, the hydrogen consumption of the fuel cell city bus is lower than 8.5 kg/100 km based on the national standard for hydrogen consumption test, which has been the top best of world. During the twelfth five-year plan, the manufacture cost of the fuel cell vehicle has been 30% lower than that of the last five-year plan, the average fault mileage is longer than 2,880 km, the mileage has reached 430 km and the hydrogen consumption is just 7.8 kg/100 km. Besides, technology of the hydrogen refueling station is mature for 35 MPa hydrogen tank. Three critical apparatus can be manufactured by domestic corporations, including the 45 MPa large volume hydrogen storage tank, the 35 MPa hydrogen refueling machine and the 45 MPa diaphragm compressor. The specific power density has reached 1.5 kW/L, the durability has reached 3,000 h and the fuel cell system can start up with environment temperature at −20 °C [11].

SPD means the specific power density. NP means the nominal power. CS means the cold storage. CSU means the cold startup. BP means the bipolar plates. GM means the General Motor Corporation.

However, when comparing the fuel cell system released by domestic corporations and that released by overseas corporations, the great gap exists in terms of performance, cost, durability, specific power density and capability of subzero startup of the fuel cell system. Parameters of some representative fuel cell engines are listed (Table 1.1) for reference.

1.1.2 Research Thesis

(1) Introduction to principle of the hydrogen fuel cell

The principle of the fuel cell is presented (Fig. 1.4). It is composed of the bipolar plates (BP), the cathode gas diffusion layer (CGDL), the cathode catalyst layer (CCL), the polymer electrolyte membrane (PEM), the anode catalyst layer (ACL) and the anode

Fig. 1.4 Principle of the hydrogen fuel cell

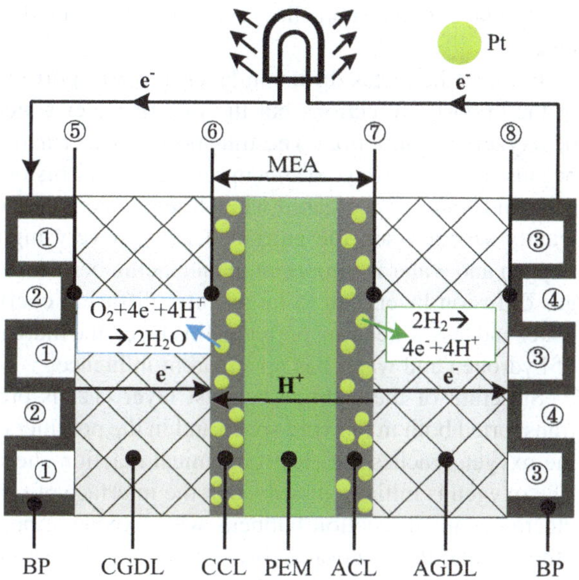

gas diffusion layer (AGDL). The ① zone is the cathode air flow channel, the ② and ④ zones the coolant flow channel, the ③ zone the anode hydrogen flow channel, the ⑤ boundary the interface of the BP and CGDL, the ⑥ boundary the interface of the CGDL and CCL, the ⑦ boundary the interface of the ACL and AGDL and the ⑧ boundary the interface of the AGDL and BP. The anode catalyst layer, the cathode catalyst layer and the polymer electrolyte membrane altogether are called the membrane electrolyte assembly (MEA).

Water plays an important role in the fuel cell and it will be elucidated as follows. To start with, humidified air is supplied to the cathode air flow channel and water vapor in the cathode flow channel is transported from or to the cathode gas diffusion layer in ways of convection transport and concentration-driven diffusion. Similarly, humidified hydrogen is supplied to the anode hydrogen flow channel and water vapor in the anode flow channel is transported from or to the anode gas diffusion layer in ways of convection transport and concentration-driven diffusion. Under different operating conditions like temperature and current density, liquid water may emerge in downstream of each flow channel and as a result, the pressure drop through the flow channel is increased.

Furthermore, the cathode gas diffusion layer and the anode gas diffusion layer are porous media and their solid state structure provides mechanical support for the MEA and path for electronic conduction. Their void space provides path for mass transport. The cathode diffusion layer is the only way for oxygen transport from cathode flow channel to cathode catalyst layer and the only way for water vapor transport from cathode catalyst layer to cathode flow channel. Similarly, the anode gas diffusion layer is the only way for hydrogen transport from anode flow channel to the anode catalyst layer and the only way for water vapor transport from anode catalyst layer to the anode flow channel or from anode flow channel to anode catalyst layer.

In the cathode gas diffusion layer, the transport of water and that of oxygen are in the opposite directions but the rate of water generation is definitely twice that of oxygen consumption. The transport of water and oxygen is mutually restricted because the effective void area in cathode gas diffusion layer is constant when there's no liquid water or decreased when liquid water is increased. This situation is significantly severe when the current density is very high and improving transport of oxygen and water is important in enhancing performance of a fuel cell. In the anode gas diffusion layer, due to larger inter-diffusion coefficient between hydrogen and water and less water transported in this layer, the mutual restriction between transport of hydrogen and water has no apparent influence.

Structure of the cathode catalyst layer [12] is presented (Fig. 1.5). Oxygen is transported both in the void space and in the polymer electrolyte solution and finally the oxygen reaches the electrochemical reaction site in the dissolved oxygen form. The oxygen is initially dissolved at the interface of solution and the void space. The electrochemical reaction happens where oxygen, proton and electronic meet each other around the solution-covered platinum. Power, heat and water are generated at the same time. Water leaves the catalyst layer either in vapor form or in liquid form or both together and enters cathode gas diffusion layer.

In the polymer electrolyte membrane [13], water and the perfluorosulfonic acid together provides path for proton conduction (Fig. 1.6). Water is transported driven by electro-osmotic drag and back diffusion [14, 15]. The electro-osmotic drag originates from combination of the water molecule and proton [16] and the average number of water molecule per proton is affected by local water content. The back diffusion originates from the gradient of local water content concentration and of course, the local water content at the interface of membrane and cathode catalyst layer is different from that at the interface of membrane and anode catalyst layer.

In the anode catalyst layer, the hydrogen molecule is ionized into proton in the polymer electrolyte solution and electronic in the solid-state electrode. The proton

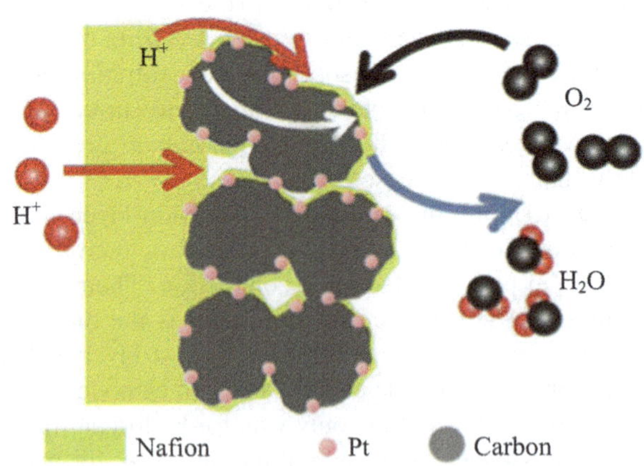

Fig. 1.5 Structure of the cathode catalyst layer [12]

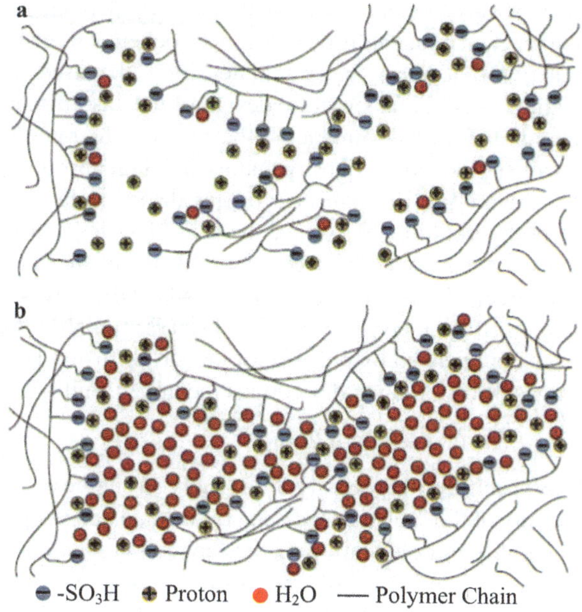

Fig. 1.6 Principle of the polymer electrolyte membrane [13]

is conducted from anode catalyst layer to the membrane and finally conducted to the electrochemical reaction site in cathode catalyst layer. To be noticed is that proton can only be conducted in solution. The proton conduction results in voltage loss usually called the ohm activation voltage. The proton conductivity is determined by temperature, local water content concentration and local proton concentration. The electronic can only be conducted in the solid-state of two gas diffusion layers and two catalyst layers. At the interface of bipolar plates and gas diffusion layers and the interface of gas diffusion layers and catalyst layers, the contact resistance emerges. The higher temperature and the larger contact pressure, the lower the contact resistance.

At the interface of gas diffusion layers and catalyst layers, the current is in completely in the form of electronic conduction. In the cathode catalyst layer, along the direction of electronic conduction, the current in the form of proton conduction is increased and that of electronic conduction is decreased. On the contrary, in the anode catalyst layer, along the direction of electronic conduction, the current in the form of proton conduction is decreased and that of electronic conduction is increased. It is definite that at any position along the proton conduction direction, the sum of the current in the form of proton conduction and the current in the form of electronic conduction are equal to output current of the fuel cell without consideration of local equilibrium and other side reaction. Usually the voltage loss in the anode catalyst layer is neglected because of the fast electrochemical rate and low activation overvoltage in comparison with that in the cathode catalyst layer.

In a word, water is byproduct of electrochemical reaction and it in turn influences the reactant concentration at the electrochemical reaction site as well as the effective reaction area. When the current density is increased, the influence is more significant so the material and structure design of the gas diffusion layer, catalyst layer and the membrane is deterministic. To be specific, the rate of water generation is relatively small at low current density and to maintain appropriate water content in the MEA, the material at the interface of gas diffusion layer and catalyst layer is required to be hydrophilic. However, the rate of water generation is large enough at high current density and the urgent target is to enhance the capability of removing water accumulated in gas diffusion layer and catalyst layer, so the material at the interface of gas diffusion layer and catalyst layer is required to be hydrophobic. Water removal is also influenced by porosity of the gas diffusion layer but the mechanical strength of gas diffusion layer determines the upper limit of its porosity.

(2) The first research thesis

The specific power density of the fuel cell engine is determined by the specific power density of the fuel cell stack and by volume and weight of auxiliary subsystems. The capability of cold startup of the fuel cell engine depends on material and structure of the fuel cell and integration of end plate with other auxiliaries. The durability of the fuel cell stack, the operating condition and the internal state of the fuel cell together govern the durability of the fuel cell engine. Optimizing design of the fuel cell stack, optimizing configuration of the fuel cell system and optimizing operating conditions of the fuel cell stack are definitely critical and the key to these optimization is monitoring and regulating water content of the individual cell and the stack.

Research [17] shows that if the fuel cell is in flooding state with too much water, the output voltage of the fuel cell is decreased with constant output current (Fig. 1.7). If the fuel cell is in drying-out state with too little water, the output voltage of the

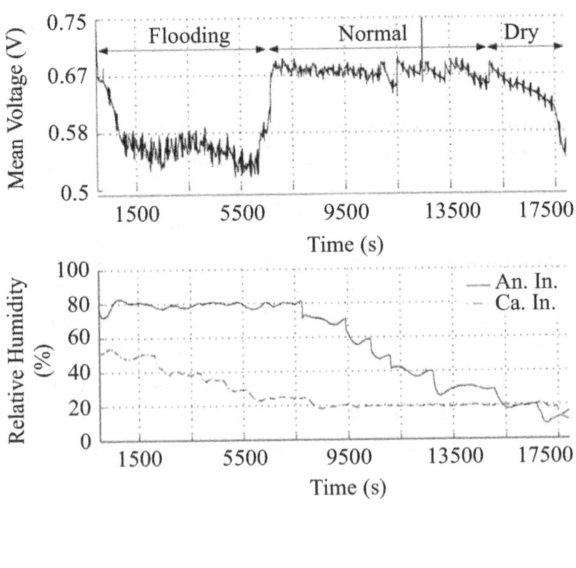

Fig. 1.7 Fuel cell performance and relative humidity

fuel cell is also decreased with constant output current. The flooding state and the drying-out state are two states for qualitative description of water content in the fuel cell and their representative characteristic is deterioration of performance of a fuel cell but it's difficult to distinguish the differencc between these two states if just based on voltage.

Measuring water content in a fuel cell is not feasible with normal approaches like the sensor for relative humidity and the weight meter, due to the coupling of complex multi-physics process. Model for water transport in the fuel cell has been proposed by scholars all over the world based on principle of the fuel cell and the known governing equation for physio-chemistry process. The model is partially validated with operating conditions and performance of a fuel cell but applying the model to various situations is difficult due to complexity of quantifying all parameters and coefficients. The analytical solution to the model is usually not obtainable and the numerical solution requires lots of calculation efforts. More importantly, the model accuracy is not guaranteed during dynamic operation of the fuel cell on its own. Therefore, we must find other ways to improve estimation accuracy of water content in both stable state and dynamic operation in combination with the mathematical model.

The AC impedance technique has been widely used in studying fundamental characteristic of the electrochemical reaction. This technique measures the AC impedance of an electrochemical device consuming power or generating power and the AC impedance depends on the frequency. In the fuel cell, the conduction of electronic and proton combines all structures together and the electronic conductivity is relatively stable but the proton conductivity is determined by water in the fuel cell. The electronic conduction and proton conduction result in ohm resistor. In the catalyst layer, the double-layer capacitor and the Faradic resistor in parallel at the three-phase reaction site result in charge–discharge characteristic and the intrinsic energy loss of the fuel cell. The double-layer capacitor and the Faradic resistor are influenced by water content in the catalyst layer. The inherent capacitor and resistor as well as parasite inductor composes the AC impedance of the fuel cell. The variation and distribution of water content shall affect the AC impedance of the fuel cell.

The impedance spectroscopy of a fuel cell [17] is plotted in the Nyquist plot and usually the higher the frequency, the smaller amplitude of the impedance (Fig. 1.8). When the fuel cell is in flooding state, the intercept of the impedance spectroscopy with the real-axis at high frequency is almost constant while the curvature of the impedance spectroscopy from low to high frequency is decreased apparently in comparison with those of a fuel cell in proper state of water content. When the fuel cell is in drying-out state, the intercept of the impedance spectroscopy with the real-axis at high frequency is increased while the curvature of the impedance spectroscopy from low to high frequency is decreased slightly in comparison with those of a fuel cell in proper state of water content.

The trend of the impedance spectroscopy of a fuel cell with varied water content is completely different for the flooding state and the drying-out state. The AC impedance would be an effective feedback for water content estimation. However, the water content cannot be calculated directly based on the AC impedance. In hence,

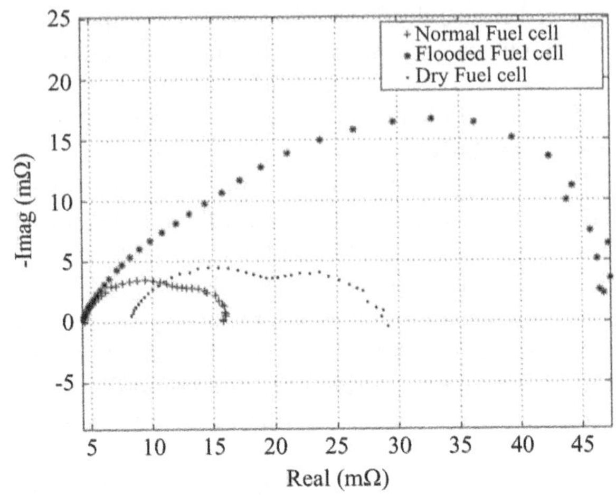

Fig. 1.8 EIS of the fuel cell and operating state

our first research thesis is to figure out the connection between water content, AC impedance and the multi-physics model of the fuel cell.

(3) The second research thesis

Based on introduction to the first research thesis, we find that the AC impedance is an effective feedback for water content estimation but measuring AC impedance of a fuel cell online is challenging. For example, when the fuel cell is applied to drive a vehicle, the difficulty of measuring the AC impedance is significantly increased due to bad environment condition and continuous electromagnetic noise and vibration.

The measurement of AC impedance should satisfy three fundamental conditions, namely the causality, linearity and stability [18, 19]. The causality means that the fuel cell presents a current or voltage response only when the voltage or current excitation signal is injected to the fuel cell correspondingly. The linearity means that amplitude of the voltage or current excitation and that of the current or voltage response must be limited. Usually, amplitude of the current excitation or response is within 10% of the stable output current while amplitude of the voltage excitation or response is suggested to be lower than the thermal voltage [20]. In practical application. Amplitude of the current excitation or response is set as 5% of stable output current [21–23] or may be smaller. The stability means that the fuel cell can operate relatively stable when the excitation is injected to it and the fuel cell can recover the stable state when the excitation is cancelled.

The commercialized electrochemical workstation in market is usually equipped in a research laboratory for the fuel cell, with wide range of measurable frequency and high accuracy of measured impedance. Specifications for some electrochemical workstation, by domestic and overseas corporation, are listed (Table 1.2). Of course, the station price is related to the nominal voltage range of the allowed device. The workstation with high nominal voltage tends to operate in current excitation mode and the electronic load is required for accurate current control, while that with low

Table 1.2 Parameters of typical electrochemical workstation all over the world

Index	KFM 2150	Model 1260A	Zennium	CHI600E
Freq. (Hz)	10 m–20 k	10 μ–32 M	10 μ–4 M	10 μ–1 M
Voltage (V)	0–150	0–46	0–10	0–10
Current (A)	660	0–100 m	0–2.5	0–250 m
Imp. (Ω)	0.0001 m–9.9999	10 μ–100 k	30 μ–1 G for CC or 1 m–1 G for CV	–
Excitation	Current	Current or voltage	Current or voltage	Voltage
Price	6 Million Yen	30 k€	–	30,000 RMB

CC means the constant current mode. CV means the constant voltage mode. Freq. means the frequency. Imp. means the impedance. The price is based on our request several years ago

nominal voltage tends to operate in voltage excitation mode and there is no special requirement for the electronic load.

The principle of workstation by KIKUSUI Corporation is illustrated (Fig. 1.9). The device KFM2150 controls the electronic load to superimpose the current excitation on stable output current of the fuel cell and the stable output current is also controlled by this device. The device KFM2151 measures voltage of each cell and the stack. The device KFM2150, KFM2151 and the electronic load together realize the synchronous sampling of voltage response and current excitation and then the AC impedance is calculated.

Through analysis, the conclusion is that measuring AC impedance has three key steps, including generating current or voltage excitation signal, synchronously extracting the voltage and current signal from stable output and fast calculation of impedance. The workstation cannot be applied to a fuel cell system in vehicle due to system cost, volume and limited hardware channel for each cell and son on.

In vehicle, the fuel cell system in combination with other power sources like lithium-ion battery pack provide power to the traction motor. There are various types of topology of the powertrain system in reference but only two types are common for vehicle application. The two types both include the fuel cell system, traction motor with inverter, the battery pack and DC/DC converter. The fuel cell system is equipped with a cell voltage monitor to monitor the voltage of each cell. Either input or output of the DC/DC converter is connected to fuel cell stack for voltage regulation and

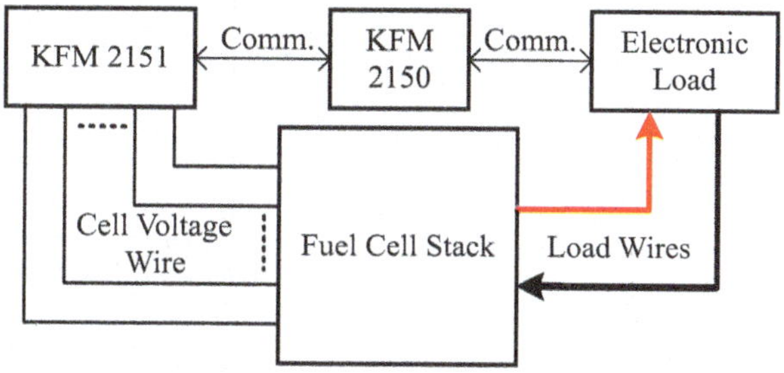

Fig. 1.9 Principle of electrochemical workstation by KIKUSUI

power match. In hence, our second research thesis is to put forward vehicle-oriented approach to measuring AC impedance online of the individual cell and the fuel cell stack based on principle of AC impedance measurement.

(4) The third research thesis

The multi-physics model of a fuel cell in terms of water transport and the online AC impedance measurement are all aimed at estimating water content of a fuel cell online. However, the measurable frequency range is limited for the online approach and measuring the AC impedance at wide frequency range takes a lot of time, which makes the system complex in function and system response slow down. The power-train system has its own intrinsic frequency and if the AC impedance is measured around such frequency point, the powertrain system may experience resonance, which weakens reliability of the system. It is necessary to reduce the target frequency point.

According to the relation between water content and impedance spectroscopy, the resistance at low frequency is affected by the water content significantly but it takes longer time to measure this resistance. The resistance at high frequency, especially the intercept with the real-axis, reflects the water content more directly. Even though variation of the resistance at high frequency is smaller in comparison with that at low frequency, it takes shorter time to measure this resistance. The representative research by Toyota Motor Corporation [24] is shown (Fig. 1.10). They measure water content of a fuel cell and the corresponding resistance of the polymer electrolyte membrane (PEM) at 1 kHz. Besides, the AC impedance at 300 Hz is very close to that at 1 kHz and the target frequency is set as 300 Hz for online AC impedance measurement.

Based on principle of a fuel cell, the transport of oxygen in flow channels and the gas diffusion layer is slow with time constant around 0.1 s while the time constant of transport of water is much longer. If the current excitation at high frequency like 300 Hz is injected to the fuel cell, the dynamic response of the fuel cell is determined by proton conduction and charge–discharge of the double-layer capacitor in parallel with the Faradic resistor in the MEA. Within milliseconds of current excitation, the

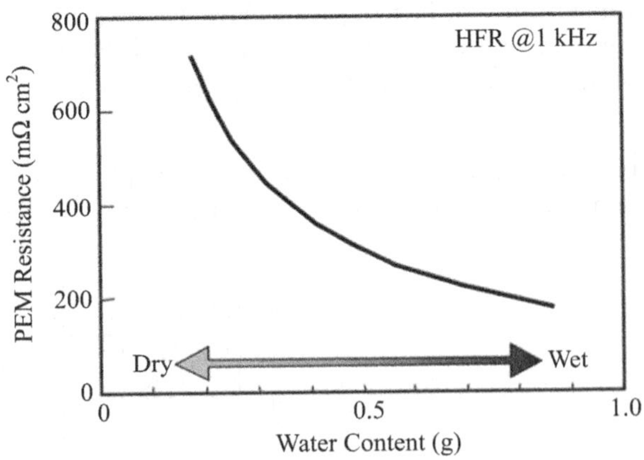

Fig. 1.10 PEM resistance and water content of the fuel cell [24]

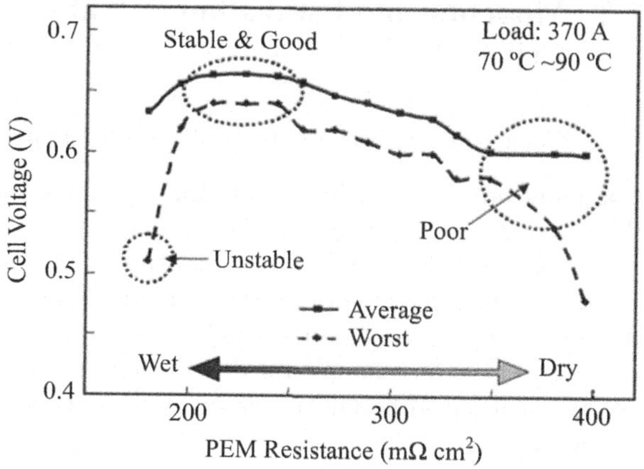

Fig. 1.11 Fuel cell performance and PEM resistance [24]

mass transport in flow channels and gas diffusion layer won't be affected and the multi-physics model of the fuel cell can be simplified to the multi-physics model of the MEA in terms of water transport. The water content estimation online is more feasible.

The Toyota Motor Corporation also measures output voltage of a fuel cell versus different water content and their relation is plotted (Fig. 1.11). When the PEM resistance is lower than 200 mΩcm^2, the average cell voltage is decreased with increased water content but the lowest cell voltage of the stack tends to be unstable. When the PEM resistance ranges from 200 mΩcm^2 to 250 mΩcm^2, the average cell voltage and the lowest cell voltage are stable around the maximal of the fuel cell. When the PEM resistance is larger than 350 mΩcm^2, the average cell voltage tends to be stable but the lowest cell voltage is decreased significantly. It is certain that too much or too little water leads to deterioration of performance of fuel cell. Proper water content guarantees the optimized performance.

The online estimation of water content provides the critical feedback for water management of the individual cell and the fuel cell stack. The fuel cell system is complex with many auxiliaries and the system control is constrained by the target performance of the fuel cell stack and that of the fuel cell system. In hence, our third research thesis is to design water management strategy for the fuel cell system to realize coordination control of subsystems and the closed-loop control of water content.

1.2 Modeling and Estimation of Water Content in a Fuel Cell

(1) Water content based on multi-physics modeling

The multi-physics modeling of the fuel cell includes the one-dimensional model, the two-dimensional model and the three-dimensional model. The one-dimensional model is mainly focused on mass transport and physio-chemistry process of the fuel cell in the direction of proton conduction. Other-dimensional model studies the distribution of mass components and physio-chemistry process through and in plane of the fuel cell, for example the distribution of water vapor and liquid water in all space and the distribution of current density in plane. They will be presented as follows.

a. One-dimensional model

Springer proposes the representative one-dimensional model of the fuel cell in steady-state and with homogeneity of temperature and gas pressure [25]. Through experiment, the relation is quantified between the water activity, water diffusion coefficient, electro-osmotic drag coefficient, proton conductivity, the local water content parameter and temperature inside MEA. The Stephan-Maxwell multicomponent diffusion equation is utilized to describe transport of oxygen, nitrogen and water vapor and transport of water vapor and hydrogen in porous media. The concentration of oxygen, nitrogen and water vapor is average of that at the cathode inlet and that at the cathode outlet. Similarly, the concentration of hydrogen and water vapor is average of that at the anode inlet and that at the anode outlet.

In the model, it's assumed that water is transported in the water vapor form in the gas diffusion layer to adapt to possible emerging of liquid water inside. Thickness of the polymer electrolyte membrane is several tens of micrometers while that of the catalyst layer is just around ten micrometers, so that two catalyst layers are considered to be part of the membrane and the activation overvoltage emerges at the interface of cathode gas diffusion layer and cathode catalyst layer and at the interface of anode gas diffusion layer and anode catalyst layer. The model is validated with good agreement with experiment result. For example, the net water flux coefficient through the fuel cell plane is significantly smaller than the electro-osmotic drag coefficient when the membrane is sufficiently humidified. When the current density is increased, the proton conduction resistance through the MEA is increased as well and it's necessary to manufacture the membrane thinner.

Springer also studies the modeling of sufficiently humidified fuel cell and experiment is performed to help correct the model [26]. Besides, the carbon-based catalyst layer is emulated by the silicone polymer based catalyst layer with the same processing craft [27]. The comparison experiment shows that the proton conductivity in the catalyst layer is far smaller than that in the membrane.

Eikerling builds a one-dimensional steady-state low-temperature model of the catalyst layer of the fuel cell to study kinetics of oxygen reduction reaction at the platinum site, proton conduction in the complex polymer electrolyte network, and oxygen transport in void space [28]. It's assumed that water exists only in the water vapor form. The analytical solution is given for four situations, like when the dynamic of oxygen transport can be neglected. The simulation result tells that the characteristic thickness of the catalyst layer is an effective indicator for penetration thickness of the oxygen reduction reaction and it's related to diffusion coefficient and local partial pressure of oxygen. The characteristic thickness is decreased when the activation overvoltage is increased.

Based on the steady-state model, the dynamic model of the catalyst layer is derived by Eikerling to study effect of kinetics of oxygen reduction reaction, double-layer capacitor, proton conductivity in the polymer electrolyte and oxygen transport in void space on dynamics of the catalyst layer [29]. The transmission line model is adopted as the equivalent circuit model of the fuel cell in the finite element analysis way. The relation between the equivalent circuit model and the physio-chemistry process is depicted.

Huang establishes the one-dimensional model for solid state electrolyte electrochemical device including the lithium-ion battery and the polymer electrolyte membrane fuel cell [30]. Much physio-chemistry process with homogeneity of temperature and solution concentration is considered, including the transport of dissolved oxygen, the charge conservation of the electrolyte solution, the electronic conduction in solid state phase, and the local potential difference between the solid state phase and the liquid phase. The research is focused on equilibrium at the open circuit voltage of the fuel cell.

Michael sets up the one-dimensional model for mass transport in the catalyst layer and the analytical solution is derived to quantify the effectively utilized thickness of the catalyst layer [31]. Water exists only in the water vapor form. The simulation result shows that the limit current density of a fuel cell is closely related to mass transport. If the oxygen and water transport is fast enough, the limit current density is proportional to thickness of catalyst layer. The model also quantifies the influence of oxygen concentration on activation overvoltage.

Raymond puts forward a one-dimensional steady-state model for the cathode gas diffusion layer and the cathode catalyst layer and the tortuous micro-pore in the catalyst layer is simplified to be a cylinder micro-pore [32]. The liquid water film forms on wall of the micro-pore and its thickness is assumed to be constant. The net water flux through the membrane is not considered. The sources of voltage loss including the activation overvoltage, proton-conduction overvoltage and electronic-conduction overvoltage are analyzed with respect to current density, which contributes to optimization of micro-structure of the catalyst layer. The effectively utilized thickness of the catalyst layer is also analyzed and the higher current density, the smaller effective thickness.

Marr aims at optimizing structure of the catalyst layer and the total amount of platinum by modeling mass transport and electrochemical reaction inside the catalyst layer [33]. The catalyst layer is full of liquid water and the simulation result shows that the effectively utilized thickness of the catalyst layer is decreased significantly with the increased current density. The content of polymer electrolyte inside gap between the carbon substrate agglomerate can be optimized to be more efficient. In fact, accumulation of liquid water in the catalyst layer must be avoided and that's where efforts are made.

Siegel investigates two-phase flow in the cathode and anode gas diffusion layer based on a one-dimensional model [34]. In the zone where liquid water accumulates, water is transported towards lower concentration by means of variation of water volume in the local micro-pore. In the zone where only water vapor exists, water is transported towards lower concentration by means of partial pressure in the local micro-pore. The interface between two aforementioned zones is where the water vapor is saturated and it's called the water front. The movement of the water front is predicted in terms of operating conditions and water content in the membrane. The water front in the cathode gas diffusion layer, that in the anode gas diffusion layer and water content in the membrane are the three indexes for determination of flooding or drying out of the fuel cell. Their work is inspiring and it's the first time that the analytical solution of the two-phase flow is presented. Please refer to the paper [35] for more research about one-dimensional model of the fuel cell.

b. Multi-dimensional model

Wang brings forward a multi-phase mixture model for multiphase multicomponent transport in capillary porous media [36]. In this model, liquid water is assumed to be dispersed water droplet in the void space like the mist and it shows the same characteristic as the water vapor, including dynamic viscosity and inter-diffusion between components. Based on conservation of mass and momentum, the dynamic model is presented for each component such as water, oxygen and nitrogen to calculate the local distribution. The classical multi-phase model is greatly simplified in terms of calculation efforts. The model is applied to study mass transport for emerging of certain substance within underground micro-pore and the simulation result shows that the calculation effort is reduced than that of the classical model but the model accuracy is guaranteed with good agreement of local concentration of the substance versus time in water [37]. Interestingly, when the emerging substance is in the bubble form, the mass transport is greatly enhanced in comparison with that of the dissolved substance and this will lead to large-area pollution.

Based on Wang's research, Um sets up a dynamic two-dimensional model of the fuel cell, taking kinetics of electrochemical reaction, distribution of current density, hydrodynamics and multi-component diffusion into consideration [38]. Besides, the dynamic three-dimensional model of fluid dynamics for inter-digitated fuel cell is presented to help understand air transport and the influence of structure design on mass transport and electrochemical reaction [39]. Furthermore, the model for MEA is introduced into the dynamic three-dimensional model and comparison is made between the fuel cell with straight flow channels and that with inter-digitated flow channels [40].

Based on Wang's research, Meng establishes a three-dimensional model with single phase of water and homogeneity of temperature to study the electronic conduction in the gas diffusion layer, that in the catalyst layer and that in the current collection plate [41, 42]. It proves that the electronic conduction resistance exists in plane of each layer. Wang proposes a three-dimensional model to investigate dynamic response of a fuel cell [43, 44]. The simulation result shows that the time constant of humidification of membrane is 10 s, that of mass transport ranges from 0.01 s to 0.1 s and that of charge–discharge of the double layer can be neglected compared with the former.

Nguyen uses a two-dimensional model to study water management, heat management and efficiency of various humidifying approaches [45]. The electro-osmotic drag and back-diffusion of water in the membrane and heat transfer from the sold state phase to the liquid phase is considered in the direction perpendicular to the fuel cell plane. Along the direction of flow channel, the latent heat resulting from evaporation and condensation of water is taken in consideration. Based on Nguyen's research, Yi studies mass transport of the interdigitated fuel cell based on a steady-state multi-component diffusion model and convection diffusion of water is introduced to describe water transport driven by the pressure gradient [46, 47].

Kazim simplifies the two-dimensional model of mass transport and compares the performance of the interdigitated fuel cell and that of the fuel cell with parallel flow channels with respect to high current density [48]. Fuller proposes a two-dimensional model of the solid-state electrolyte fuel cell and the theory of concentrated solution is used to explain electrochemical process in the membrane [49]. The focus is on water management, heat management and the efficiency of hydrogen utilization. Similarly, West uses the theory of concentrated solution with variable transport characteristic to calculate local potential and local water content [50]. The land of the bipolar plates provides mechanical support for the gas diffusion layer and it also limits hydrogen and oxygen to reach reaction sites between two lands on either side of the MEA. He tries to study the effect of the bipolar plates land on performance of a fuel cell. Please refer to papers for more research of multi-dimensional model [51, 52] and papers for modeling of two-phase flow [53–56].

(2) Water content based on semi-empirical model

Maggio studies the water transport based on semi-empirical equation and the focus is on the relation between current density and porosity of gas diffusion layer [57, 58]. It's predicted that part of the polymer electrolyte membrane which is close to the anode catalyst layer tends to be dried out. The model can predict the ohm overvoltage and activation overvoltage under the constant operating condition and they are validated with experiment on a fuel cell with good accuracy.

McKay sets up the dynamic lumped model of the water concentration in anode chamber and cathode chamber of a fuel cell in open-circuit state [59]. An open-loop nonlinear state observer is designed for water content in the polymer electrolyte membrane and input parameters of the observer include pressure and temperature of the gas at inlet and outlet of both cathode and anode, the relative humidity of supplied gas at inlet of both cathode and anode. Based on experimental calibration, the observer is capable of quantifying water transport through the membrane for 24 cells.

Jiao puts forward a three-dimensional model for cold startup behavior of a fuel cell and aspects are considered in the model, including freezing of liquid water in membrane, unbalanced water transport at interface of catalyst layer and gas diffusion layer, and freezing and melting of liquid water in catalyst layer and gas diffusion layer [60]. Through comparison of successful and failed cold startup, they find that increasing the content of polymer electrolyte in the catalyst layer contributes more to reducing the freezing of water than increasing thickness of the polymer electrolyte membrane.

Wang builds a dynamic three-dimensional model of a fuel cell to analyze the adsorbing and transport of water in the membrane and the catalyst layer at subzero temperature [61]. It's found that the initial water content in the MEA is the key to successful cold startup. There exists the critical water content for cold startup from different initial temperature. Excessive water in the catalyst layer will freeze before the fuel cell reaches temperature at 0 °C and the void space for oxygen transport will be blocked. Please refer to the paper [62–66] for research about cold startup and the paper [67–69] for research about purging water for cold storage.

(3) Water content based on experiment equipment

The practical distribution of water content in the fuel cell is unknown without proper sensors to measure local relative humidity of water vapor and local content of liquid water and the accuracy of multi-physics model cannot be guaranteed without support of exact experiment data [70]. Scholars have tried to analyze water content based on available experiment result and summarize the empirical governing equation for water content related to operating condition.

The optical equipment is applied to study water content. For example, Ous studies the relation between operating condition and water content in the flow channel [71]. For the flow channel, raising operating temperature can enhance evaporation of liquid water and the stoichiometric ratio of supplied air and hydrogen has the similar effect on accumulation of liquid water. Increasing flow rate of supplied air contributes to removing liquid water in the flow channel. The formation of liquid water in the flow channel is related to current density.

Liu investigates the effect of operating condition, including pressure drop of cathode chamber and that of anode chamber, on liquid water in the flow channel [72, 73]. The pressure drop is determined by liquid water content and the pressure drop of cathode chamber is higher than that of anode chamber due to no liquid water in the anode flow channel. The emerging of liquid water in the flow channel leads to deterioration of performance of a fuel cell.

Yang tries to figure out the mechanism of formation of water droplet on surface of the gas diffusion layer and the characteristic of mixed transport of water vapor and liquid water [74, 75]. It's found that water droplet only emerges on certain position of the surface and if the material of the gas diffusion layer is hydrophilic, the film of liquid water forms much easier on the surface. Liquid water tends to accumulate around corner of the flow channel.

The neutron-radiography technology is adopted to study water content as well. For example, Mosdale quantifies the distribution of water content in the polymer electrolyte membrane along the direction of proton conduction using the neutron-radiography technology [76]. The relative humidity of supplied hydrogen determines the trend of distribution of local water content in the membrane. Even when the content in the membrane is very low, the membrane can still support the proton conduction. Bellows measures the distribution of water content in the membrane with 500 μm thickness and no apparent gradient of local water content is observed [77]. Researchers also try to quantify the water content and they pay attention to the effect of flow channel structure and current density on distribution of water content in the fuel cell [78–88].

The pressure and temperature sensor are also utilized to study water content. For example, Steiner trains the fuel cell model based on pressure drop of the cathode chamber and the output voltage [89]. The actual operating state of the fuel cell is compared with that calculated by the neural network algorithm and their difference is set as the threshold for determining state-of-health of the fuel cell. Their research is partially validated with experiment. Ito investigates the relation between output voltage of a fuel cell and the saturation of water vapor [90]. The experiment is conducted on a fuel cell with inter-digitated flow channel and the AC impedance and the pressure drop of both cathode chamber and anode chamber are measured as well. The flooding of the fuel cell can be identified exactly.

Nguyen studies the relation between water content and pressure drop of both cathode and anode chamber [91–94]. The pressure drop is the direct evidence for flooding with liquid water accumulating in the flow channel. Increasing the flow rate of supplied gas and the operating temperature can effectively remove liquid water. Pei investigates the pressure drop of anode chamber with respect to water content in the fuel cell [95, 96]. When the internal water content of the fuel cell changes from the appropriate to the excessive, the variation trend of the pressure drop shows the alternation of two stages and two slopes versus time. The mechanism for the pressure drop is established [97, 98].

Li studies the pressure drop of cathode chamber with respect to water content in the fuel cell [99]. When the internal water content of the fuel cell changes from the appropriate to the excessive, the variation trend of the pressure drop is very clear and reproducible and based on this, the flooding of a fuel cell can be identified. Haluk proposes the water content estimation of a fuel cell in drying-out state [100, 101]. The relation between resistance and water content of the polymer electrolyte membrane is quantified based on the model in reference and data by pressure and temperature sensors. The experiment validates the accuracy of the method.

In hence, the pressure drop of the cathode chamber is the efficient evidence for flooding state of the fuel cell. However, applying this index to the fuel cell system in operation is challenging. Firstly, the accumulation of liquid water emerges in the catalyst layer at first and then it diffuses to the gas diffusion layer and the flow channel. The obvious pressure drop shows only when liquid water emerges in the flow channel and in this case, the fuel cell has been in severe flooding state. Secondly, the effective surface area of the fuel cell in vehicle is very large. The drying-out, normal and flooding state exist at the same time in plane of the fuel cell at high current density and low relative humidity of supplied air and it's difficult to quantify the flooding state of the fuel cell in this situation. Thirdly, when hundreds of fuel cells are assembled in serial, pressure sensors and temperature sensors are installed only at inlet and outlet of both cathode and anode chamber. Due to difference in pressure and temperature of each cell, it's difficult to identify the flooding state of each cell.

(4) Water content based on AC impedance

The effect of operating condition on performance of a fuel cell is reflected on the electrochemical impedance spectroscopy and under inappropriate operating condition, water content of a fuel cell will be away from the equilibrium state. The equivalent circuit model is an approximation of the impedance spectroscopy and through experiment the sensitivity of model parameters to operating condition can be analyzed to quantify their relation. Usually, the AC impedance at high frequency range is related to double-layer capacitor and Faradic resistor in the three-phase reaction site. The high frequency resistance whose phase is zero is sum of proton conduction resistance and electronic conduction resistance [102]. The AC impedance at low frequency range represents the resistance of mass transport [103, 104]. This is the foundation for water content estimation based on AC impedance.

Springer analyzes how the operation condition affects impedance spectroscopy of a fuel cell [105]. The experiment result shows that the impedance is closely related to operating condition of supplied gas at the cathode. The impedance at high frequency range describes the transfer resistance of surficial charge and characteristic of the catalyst layer. Gas transport in the gas diffusion layer dominates the impedance at low frequency range, particularly the limit of mass transport. If pure oxygen instead of air is supplied to the cathode, the impedance at low frequency range is significantly reduced. If the fuel cell is not sufficiently humidified, the voltage loss emerges resulting from proton conduction in the membrane, that in the polymer electrolyte of the catalyst layer and the activation overvoltage of the electrochemical reaction. At medium current density, the proton conductivity in the polymer electrolyte of the catalyst layer and the effective diffusion coefficient of oxygen in the catalyst layer is the critical factor for variation of the impedance spectroscopy. At high current density, the mass transport affects performance of the fuel cell significantly and that is reflected by the impedance at low frequency range.

Yuan performs experiment on a fuel cell stack with six cells in serial [106, 107]. Each time after changing operation condition, the impedance spectroscopy of each cell and the stack is measured. The experiment result shows that the high frequency resistance whose phase is zero varies with current density and the Faradic resistor in parallel with the double-layer capacitor is decreased quickly at first and then increased gradually with increased current density. If the fuel cell is sufficiently humidified, the high frequency resistance whose phase is zero is almost constant while curvature of the impedance spectroscopy at high and low frequency range is decreased at the same time with increased current density. If the temperature is increased, the high frequency resistance whose phase is zero is decreased slightly and curvature of the impedance spectroscopy at high and low frequency range is increased at the same time. At the same current density, the Faradic resistor in parallel with the double-layer capacitor is decreased with increased temperature and it's reasonable.

Besides, the curvature of the impedance spectroscopy at high and low frequency range is increased at first and then decreased with increased current density, if flow rate of the supplied air is the same and the supplied air is sufficiently humidified, while the high frequency resistance whose phase is zero is stable. If humidification of the supplied hydrogen is interrupted, curvature of the impedance at low frequency range is decreased slightly and the whole impedance spectroscopy moves rightwards on the Nyquist plot, but the temperature has little influence on their behavior. On the contrary, if humidification of supplied air is interrupted, curvature of the impedance at low frequency range is decreased gradually and the whole impedance spectroscopy moves rightwards on the Nyquist plot. The higher the temperature, the more obvious the trend of their behavior.

Hou studies the impedance spectroscopy of a fuel cell stack with twenty cells in serial experimentally [108]. The test result shows that curvature of the impedance at low frequency range is decreased with decreased stoichiometric ratio of supplied air. The relative humidity of supplied air and operating temperature mainly impacts curvature of the impedance at high frequency range. If output current of the fuel cell experiences a step change, the high frequency resistance with zero phase is increased at low current density while it is decreased at high current density. It tells that the high frequency resistance with zero phase is an effective indicator for qualitative estimation of water content. Aaron [109], Malevich [110, 111] and Mathias [112] also investigate the effect of relative humidity on the impedance spectroscopy and they have similar conclusions.

Kim tries to figure out influence of clamping torque on performance of fuel cell in terms of high frequency resistance and mass transport resistance [113]. Under the same operating condition, the high frequency resistance and the net water flux through the membrane are decreased with the increased clamping torque. Asghari finds that if the clamping torque is uneven in plane of the fuel cell, the high frequency resistance is increased and the mass transport is deteriorated as well as performance of the fuel cell [114]. After the fuel cell stack is assembled, a tighter compression for short-time duration guarantees adequate water content in the MEA [115, 116]. In addition, the high frequency resistance and the charge transfer resistance are decreased with increased clamping torque.

Freire uses pure oxygen instead of air to investigate the impact of membrane and operating condition on performance of a fuel cell [117]. A thinner membrane means the better effect of water management and the sensitivity of high frequency resistance to operating temperature and current density is weakened. The impedance at low frequency range is decreased in comparison with that with supplied air and it demonstrates that nitrogen in air limits the oxygen transport in cathode gas diffusion layer and catalyst layer [103].

Oszcipok studies the impedance spectroscopy during cold startup of a fuel cell with double serpentine flow channels [118]. The high frequency resistance with zero phase is decreased at first and then tends to be stable with increased current density, while the Faradic resistor in parallel with the double-layer capacitor is increased gradually. Besides, the high frequency resistance and output voltage of the fuel cell is significantly influenced by the flow rate of supplied hydrogen.

These experiment-based research demonstrates that the AC impedance can reflect performance and internal state of water content of the fuel cell with respect to operating condition and structure design. Furthermore, researches try to emulate the impedance spectroscopy and quantify the mathematical relation between impedance spectroscopy and characteristic of a fuel cell.

Representatively, Randles derives the equivalent circuit model based on the electrochemical reaction of metal-ion in the solution [119, 120]. This model is composed of a frequency-dependent capacitor in serial with a frequency-dependent resistor as the first circuit, a capacitor in parallel with the first circuit as the second circuit, and a resistor in serial with the second circuit. Grahame derives the capacitor, depending on frequency and mass transport, from the second Fick diffusion law [121]. Fouquet improves the model [17] proposed by Randles. The double-layer capacitor is replaced with the constant phase element (CPE) to reflect the non-uniformity of physic characteristic and that of surficial current density in the catalyst layer. The frequency-dependent capacitor and resistor are replaced with a constant Faradic resistor in serial with a Warburg element to describe the semi-infinite diffusion of oxygen in catalyst layer and gas diffusion layer. In experiment the fuel cell experiences the flooding state, the normal state and the drying out state and the internal state of water content is plotted with respect to three parameters of the equivalent circuit model. The internal state of water content of the fuel cell can be differentiated clearly with obvious trend.

The transmission line model is also proposed to study characteristic of porous electrode of catalyst layer [122–126], especially the distribution of proton conduction resistance inside [124, 127]. To simplify the derivation, the void space in the catalyst layer is considered to be composed of the cylinder micro-pore with the same diameter and the electrochemical reaction site is evenly distributed in the catalyst layer [125]. Besides, the characteristic of electrochemical reaction is assumed to be homogeneous. After dividing the catalyst layer into lots of micro units, the mathematical model of the physio-chemistry process in each unit can be given with respect to the following: the potential of the solution resulting from proton conduction resistance; the potential of the solid-state electrode resulting from electronic conduction resistance; the electrochemical reaction rate along with the potential difference between the solid-state electrode and the solution. If the number of the micro unit is infinite, the transmission line model can quantify characteristic of the catalyst layer with good accuracy [126, 128]. Scholars suggests using the Gerischer element to describe the surficial electrochemical reaction and the mass transport [129] and using the corrected Fick diffusion law to describe the steady-state voltage loss and dynamics of the catalyst layer [130, 131].

In hence, the impedance spectroscopy is a helpful tool for analysis of the fuel cell because of its simple operation, high sensitivity and intuitive quantification. However, measuring the impedance spectroscopy over wide frequency range takes a lot of time [132] and the resistance at low frequency may be scattered [133]. Fitting parameters of the equivalent circuit model requires lots of calculation efforts and it is limited to linearization zone of the fuel cell around the preset operating condition [134]. If the effective surface area of the fuel cell is large, the shunt resistor must be utilized to measure output current of the fuel cell but the resistance of the shunt resistor varies with frequency and temperature due to heat dissipation [135]. Measuring the AC impedance at low frequency may lead to oscillation of oxygen concentration in the gas diffusion layer and the air supply subsystem of the fuel cell system may be impacted [136].

The deployment of test probe influences accuracy of the measured AC impedance and the common two-probe mode is very convenient with high accuracy for a fuel cell of small effective surface area [137, 138]. When measuring AC impedance of a fuel cell with large effective surface area, the position of the probe affects the result so the position must be properly chosen. Scholars try to segment the fuel cell into several sections and measure AC impedance of each section to study distribution of water content [139]. The neutron-radiography technique is also applied to help prove capability of AC impedance in estimating water content [140, 141] but in fact, segmentation will destroy the electrical integrity of a fuel cell.

1.2.1 Online AC Impedance Measurement Methods

(1) Methods based on the electronic load

Brunetto suggests using the programmable electronic load to measure the AC impedance [142, 143]. Firstly, the electronic load is controlled to draw a constant current from the fuel cell; secondly, the function generation sends command to the electronic load to superimpose current excitation on the constant current; thirdly, the fuel cell voltage and the current excitation signal are measured and the impedance is calculated based on sampled signals. The frequency of the impedance ranges from 7.6 MHz to 10 kHz. Then they perform experiment on a fuel cell stack with 16 cells in serial but the experiment data is not compared with that of the electrochemical workstation. It's hard to confirm the accuracy of this method based on their data.

Wasterlain improves the method based on the electronic load and their work is supported by the DIAPASON projected launched by Government of France [144–146]. To be specific, the PXI hardware by National Instruments Corporation replaces the function generator to control the electronic load and it is also capable of measuring the voltage and current simultaneously with high sampling speed and wide band frequency. The device can measure at most voltage of 31 cells to calculate AC impedance. The frequency of the impedance ranges from 0.05 Hz to 5 kHz. The experiment is conducted on a fuel cell stack with 3 cells in serial and the comparison is made between data based on their method and that based on the electrochemical workstation by ZAHNER Corporation. Large difference exists between two group data and their accuracy is doubtful.

Furthermore, Petrones proposes to install a DC/DC converter between the fuel cell stack and the electronic load to measure the AC impedance and their work is supported by the subproject of the fuel cell fault diagnosis based on the DC/DC converter [147, 148]. This subproject belongs to the fuel cell subproject of the 7th frame project launched by department of the Europe Union. Even though the DC/DC converter adjusts the output of the fuel cell stack, the current excitation still comes from the extra electronic load which is connected to the fuel cell stack in parallel with the DC/DC converter.

(2) Methods based on the DC/DC converter

The Plug Power Incorporation in U.S.A. lists various approaches to inject current excitation or voltage excitation signal to the fuel cell in their patent, such as the extern signal oscillator, the ripple current generator and power converter [149]. The key technology is also mentioned related to signal processing and signal sampling. The AVL Corporation in Austria advises to install an excitation signal generator in parallel to the fuel cell stack and a capacitor is connected between the fuel cell stack and the signal generator to isolate the constant direct current and lower the power consumption by this device [150–152]. Their specific embodiment is to inject current excitation to the fuel cell stack and a micro controller unit measures the voltage and current through sensors to calculated impedance. The suggested frequency range is from 8 to 500 Hz.

Narjiss introduces their AC impedance measurement based on the DC/DC converter [152–154]. The topology of this converter has a multi-stage structure with a DC/AC converter, an AC/AC converter and an AC/DC converter is serial. The micro controller controls the duty cycle of power semiconductor switching in each converter to draw a constant current from the stack and superimpose a current excitation on the stable current by adjusting the duty cycle. The frequency range and the accuracy of the impedance are not mentioned yet.

Hinaje suggests measuring the AC impedance based on ripple current of the input current of a DC/DC converter [155]. The frequency of the ripple current depends on the PWM frequency of the power semiconductor switching. Usually, the PWM frequency is higher than 10 kHz and measuring voltage and current signal at such high frequency is challenging. The accuracy of the impedance is not guaranteed. By analyzing the spectroscopy of input current of the DC/DC converter, Dotelli tries to use an electronic load to simulate the ripple current of the converter and then experiment is performed on a fuel cell [156–159]. The frequency of the impedance ranges from 100 Hz to 2 kHz with only five frequency points available.

The Samsung Electronics Corporation in Korea presents a more practical proposal in patent [160]. To be specific, the power converter is installed between the fuel cell stack and a load. Another DC/DC converter is connected to the fuel cell stack in parallel with the first converter and the output of the second converter supplies power to auxiliaries such as water pump in a fuel cell system. If so, the overall efficiency of the fuel cell system is ensured with little power loss through the converter. The second DC/DC converter superimposes current excitation on the fuel cell stack and the frequency of the excitation signal can be flexibly regulated.

Toyota Motor Corporation in Japan realizes the water content estimation based on the DC/DC converter. This proposal is applied to their fuel cell vehicle namely the FCH-adv [24, 161]. To be noticed is that the DC/DC converter is connected between the fuel cell stack and the battery pack. Firstly, the DC/DC converter is controlled to inject a voltage excitation to the fuel cell stack; secondly, the voltage of a fuel cell and the stack and the current of the stack are measured simultaneously based on sensors and the micro controller. The sampling number of each signal is 512 and the target frequency of AC impedance is 300 Hz. Based on measured signals, the impedance is calculated online for real time. In patent, the topology of the DC/DC converter and the powertrain system is elucidated [162, 163] and the method for generating the voltage excitation signal presented as well [164].

Recently, Toyota Motor Corporation has improved their powertrain system of the fuel cell vehicle [165, 166]. One more DC/DC converter is connected to the output of the fuel cell stack and this converter is integrated with the function of AC impedance measurement. The output of this DC/DC converter is connected to input of the DC/DC converter for the battery pack. The topology of the fuel cell DC/DC converter is three-phase interleaved boost converter. The method for generating the voltage excitation signal is also presented in patent in detail [167]. Their work is the most attractive until now and this is the benchmark for our research.

(3) Other methods

Hyundai Motor Corporation in Korea suggests acquiring the internal resistance of a fuel cell based on the step current signal [168]. This apparatus is composed of the cell voltage monitor, the main load and an extra power consumption device. The main load is controlled to maintain stable output of the fuel cell stack. The extra power consumption device is switched on and off at target frequency. The cell voltage monitor obtains the voltage response of a fuel cell and then the internal resistance is calculated based on certain signal processing methods. Their experiment data of a fuel cell stack is comparable to that of a standard current interruption equipment. Their method is very practical to some extent.

Nissan Motor Corporation in Japan put forwards a more compact approach to measure AC impedance online with low power consumption [169]. The key concept is that the generation of excitation signal and the measurement of excitation signal and response signal are integrated into the cell voltage monitor. To be specific, two cells with three voltage connection points are connected to the circuit for excitation signal generation. Three capacitors are connected between the three voltage connection points and the circuit input to isolate direct current. The middle connection point is shared between voltages of two cells. The circuit for excitation signal generation is composed of the amplifier, capacitor and resistor. The microprocessor controls its output voltage to inject current excitation to the two fuel cells and then the voltage and current are measured through the same circuit. The impedance is calculated by the microprocessor. Please refer to the reference for more details.

1.2.2 Water Management Strategy of the Fuel Cell System

The fuel cell system is very complex and the water content and its distribution in a fuel cell influences performance of the fuel cell. Estimation of water content provides feedback for the coordination control of the fuel cell system. The water content of a fuel cell is governed by the operating condition and the water management strategy is actually founded on dynamic regulation of the operating condition based on the feedback water content [71, 170]. The target of the strategy is to optimize performance and durability of the fuel cell stack.

Bosco analyzes experimentally the pressure drop in the anode flow channel and that in the cathode flow channel with respect to various operating conditions [171]. Based on the experiment result, the threshold of pressure drop is set for operating conditions and output current of a fuel cell to judge whether the fuel cell is flooded. When the actual pressure drop is larger than the preset threshold, some counter-flooding measures are taken to remove excessive water, like closing the humidifier, reducing the output current, increasing flow rate or lowering the operating pressure of supplied air and hydrogen. Similarly, Rodatz summarizes the relation between the cathode pressure drop and the operation condition of a fuel cell and puts forward three ways to improving the performance [172]. To be specific, the current pulse contributes to prevent the formation of oxidation layer on the platinum agglomerate, the impulse of flow rate and operating pressure of supplied air helps purge water from flow channels, and the pressure difference between the anode chamber and the cathode chamber suppresses the water flux from cathode to anode.

Song tries to find the optimization interval of the operating condition for purging water based on pressure drop experimentally [95]. The water management strategy based on regulating the cooling temperature is proposed. When the fuel cell is flooded, the cooling temperature is raised; when the fuel cell is dried out, the cooling temperature is decreased. The strategy is validated with experiment on a fuel cell stack.

Barbir finds that the increase of pressure drop in the cathode flow channel is a reliable index for flooding of a fuel cell and the high frequency resistance a reliable index for drying out of a fuel cell through massive experiment on the stack [173, 174]. During operation of the stack, by monitoring the pressure drop of the cathode flow channel and the high frequency resistance, the qualitative analysis of water content of a fuel cell is feasible and the water management strategy can be established in this way. Kurz states that the resistance at 0.5 Hz and that at 1 kHz together can diagnose the flooding or drying out of the fuel cell [175]. The resistance at 0.5 Hz is the index for flooding and the control-oriented model of water management is set up to regulate the flow rate of supplied air. By improving the control algorithm, the water management strategy is adapted to different operating conditions like temperature, current density and relative humidity.

Karnik sets up the lumped model for water transport inside the cathode and the anode chamber as well as water transport through the membrane, and the key concept is water equilibrium in the fuel cell [176]. Through simulation, the model is verified to be capable of predicting flooding and drying out of the fuel cell. The water management strategy is established based on the lumped model. Then they analyze the capability of strategy to maintain relative humidity of the supplied hydrogen when no liquid water emerges in the anode flow channel. Besides, they analyze whether water content in the fuel cell can be equilibrated by means of regulating relative humidity of supplied air and purging waste gas and water from the anode. Then they put forward the closed-loop control algorithm for water content. No experiment is performed to confirm the effective of their strategy.

Fang studies the mechanism of purge process in the anode subsystem [177]. The dynamic model of the dead-end anode subsystem and that of the hydrogen injector subsystem are proposed and analyzed. A robust model predictive control algorithm is designed to adjust the pressure at the anode inlet and the simulation result verifies the capability of this algorithm with good convergence. Then the algorithm is applied to the micro controller unit and a test plant is set up for the two subsystems. The experiment result validates the good accuracy and the fast dynamic response.

Cheng [178] and Kim [179] suggest utilizing exhausted waste gas from cathode outlet to humidify the fresh air into cathode inlet. Hu sets up the dynamic model of the air supply subsystem when the exhausted waste gas from cathode outlet is partially recirculated to be mixed with fresh air [180]. The flow rate of recycled exhausted water gas is controlled by adjusting the open angle of valve installed in the recirculation pipe. The simulation result shows that the recirculation of the exhausted waste gas can effectively control water content in a fuel cell. Zhao studies the control strategy for fuel cell system with both recirculation of exhausted waste gas from cathode and that from anode [181]. The orthogonal experiment is performed to validate capability of maintaining water content and output voltage of the fuel cell. The workstation measures the high frequency resistance and provides feedback to the control algorithm. The result shows that the recirculation of water gas can regulate output voltage of a fuel cell and decrease the difference of oxygen concentration between that at inlet and that at outlet of the cathode chamber. The more interesting result is that the fuel cell can maintain proper water content even when only dry air is supplied to the fuel cell system.

Bao puts forward a model of the fuel cell to estimate partial pressure of oxygen in the cathode chamber [182]. The neural network is used to train the model of air supply subsystem especially the air compressor model. The control algorithm of the air supply subsystem is optimized based on the genetic algorithm, in terms of the air stoichiometric ratio and gas pressure at the chamber inlet. Three types of air supply subsystem are simulated and it finds that recirculation of waste gas can improve the system efficiency.

The Toyota Motor Corporation proposes the water management strategy based on the high frequency resistance namely the PEM resistance [183]. If the fuel cell is detected to be in flooding state, the air stoichiometric ratio is increased while the gas pressure at the cathode inlet is decreased. If the fuel cell is detected to be in drying-out state, the air stoichiometric ratio is decreased while the gas pressure at the cathode inlet is increased. In the first generation of the fuel cell stack (presented for the first time in reference), water content is managed in such way (Fig. 1.12a) and the variation of water content throughout the power range is relatively large. In the second generation of the fuel cell stack (installed in the fuel cell vehicle named MIRAI), the humidifier in the air supply subsystem is cancelled but the practical fuel cell system shows stable water content throughout the power range (Fig. 1.12b).

In hence, the efficient water management must combine four parts together, including the pressure drop of the cathode chamber, the pressure drop of the anode

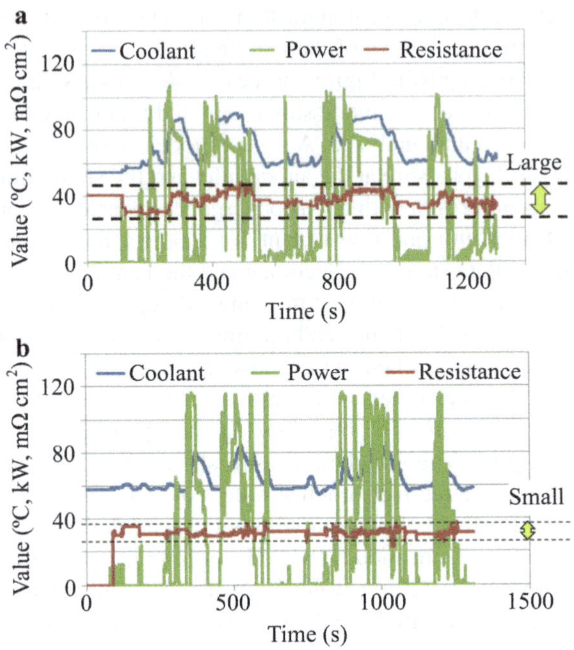

Fig. 1.12 Water management of the fuel cell system by Toyota

chamber, the high frequency resistance and the dynamic model of subsystems. This will be the guidance for our research and please refer to the paper [184–188] for more research about water management of a fuel cell system.

1.3 Introduction to Our Research Content

Our research content is presented from four aspects (Fig. 1.13), including the modeling of water content in MEA for a PEM fuel cell, the approach to online AC impedance measurement of a fuel cell and the stack, the water content estimation of a fuel cell and water management strategy for the fuel cell system.

(1) Modeling of water content in MEA for a PEM fuel cell

Firstly, the polarization curve of the fuel cell and the equivalent circuit model are analyzed to establish the relation between sources of energy loss of a fuel cell and components of the equivalent circuit model. Secondly, a steady-state one-dimensional model of a fuel cell with large surface area and special structure design is proposed to study internal recirculation of water content and its effect on proton conduction and performance of a fuel cell. Finally, a steady-state pseudo-two-dimensional model of the MEA is built to analyze the effect of water content on

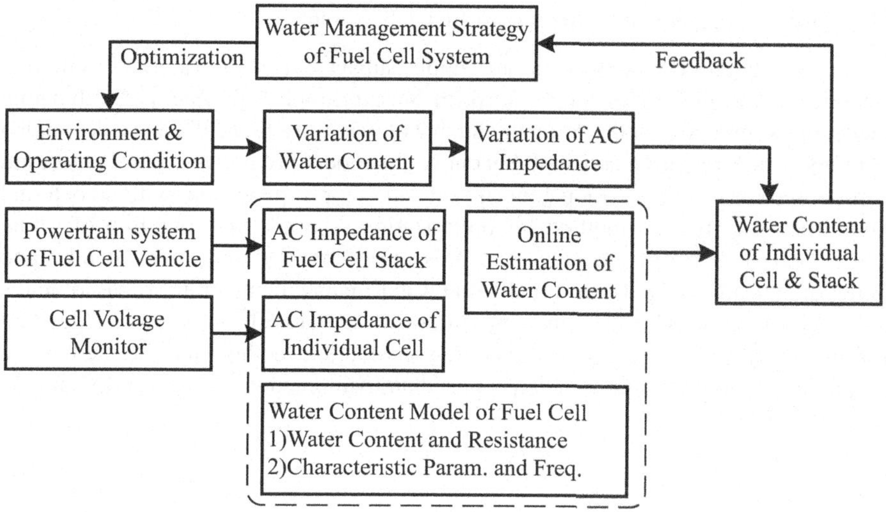

Fig. 1.13 Our research content

activation overvoltage and the ohm overvoltage and that on the equivalent circuit model.

(2) Approach to online AC impedance measurement of a fuel cell and the stack

Firstly, the topology of powertrain system of a fuel cell vehicle is analyzed and the topology for the specialized DC/DC converter is proposed with the integrated online AC impedance measurement function. Secondly, the average model and the small signal model of the DC/DC converter are derived to describe its dynamic response to the duty cycle and the transfer function for generation of current excitation against the duty cycle fluctuation is derived and verified through simulation. Thirdly, the signal processing circuits are proposed for extraction of weak signal like the voltage response of a fuel cell and the voltage drop over the shunt resistor due to current excitation, and the cell voltage monitor is redesigned to support the impedance calculation. Finally, the test plant for the designed DC/DC converter, the new cell voltage monitor and the fuel cell stack is set up to verify feasibility of the proposed approach to online AC impedance measurement.

(3) Water content estimation of a fuel cell

Firstly, the error source of the online measured AC impedance is analyzed theoretically and compensated based on calibration method, and the experiment is performed to make a comparison between the AC impedance by the electrochemical workstation and that by our approach. Secondly, the relation is analyzed between water content of a fuel cell with self-humidifying MEA and two indexes, such as the high frequency resistance and the statistics of cell voltage, are proposed. Finally, the method for water content estimation of a fuel cell with external-humidifying MEA is proposed and it's verified through simulation.

(4) Water management strategy for a fuel cell system

Firstly, the dynamic model of the air supply subsystem of the fuel cell system is established and the closed-loop control of the air supply subsystem, the dynamic model of oxygen in cathode chamber and that of water in a fuel cell can be decoupled in terms of different time constant. Secondly, the model predictive control algorithm is proposed for the closed-loop control of the air supply subsystem with no external humidifier. Thirdly, the method for optimization of operating condition of a fuel cell is presented based on the partial pressure of oxygen and water at inlet and outlet of the cathode chamber. Finally, the water management strategy, based on the approach to online AC impedance measurement, the closed-loop control algorithm for air supply subsystem, the water content estimation method and the method for optimizing operating condition, is proposed and validated through experiment on a fuel cell stack.

1.4 Organization of This Book

This book has five chapters and they are organized as follows: the first chapter introduces the background and the state-of-art related to our research; the second chapter presents the modeling of water content in MEA for a PEM fuel cell; the third chapter is focused on the approach to online AC impedance measurement of a fuel cell and the stack; the fourth chapter elucidates methods for water content estimation of a fuel cell; the fifth chapter describes the water management strategy for a fuel cell system.

References

1. International Energy Agency (2017) world energy outlook 2017. International Energy Agency, Paris
2. Ballard Power Systems. Ballard Milestones[EB/OL]. (2017–12). http://www.ballard.com
3. The DOE Hydrogen and Fuel Cells Program Plan[S/OL]. (2011–09). http://www.eere.energy.gov/hydrogenandfuelcells/program_plans.html
4. Tien N, Jake W. Well-to-wheels greenhouse gas emissions and petroleum sse for mid-size light-duty vehicles[R/OL]. (2010–10–05). http://www.hydrogen.energy.gov/h2a_analysis.html
5. McConnell VP (2007) Downsized footprint and material changes for GM's fourth-generation fuel cell technology. Fuel Cells Bull 2007(1):12–15
6. Fuel cell transit bus evaluations, joint evaluation plan for the U.S. Department of Energy and the Federal Transit Administration[R/OL]. (2010–11). http://www.nrel.gov
7. Aso S, Kizaki M, Mizuno H (2009) Development progress of the Toyota fuel cell hybrid Vehicle. SAE Int J Engines 1(1):296–303
8. Jones DJ (2013) Global change, energy issues and regulation policies: introduction to hydrogen and fuel cell technologies and their contribution to a sustainable energy future. Springer, Dordrecht, pp 161–178

9. Lim T, Ahn BK (2013) Hyundai's FCEVs: a pathway to new possibilities. ECS Trans 50(2):3–10
10. Saxe M, Folkesson A, Alvfors P (2008) Energy system analysis of the fuel cell buses operated in the project: clean urban transport for Europe. Energy 33(5):689–711
11. Xu L, Li X, Hua J (2009) Parameter identification and control strategy optimization of hybrid fuel cell powertrain. J Mech Eng 45(2):56-61
12. Kazunari S, Li H, Akari H et al (2016) Hydrogen energy and engineering: a Japanese perspective. Springer, Tokyo
13. Jiao K, Li X (2011) Water transport in polymer electrolyte membrane fuel cells. Prog Energy Combust Sci 37(3):221–291
14. Karimi G, Li X (2005) Electroosmotic flow through polymer electrolyte membranes in PEM fuel cells. J Power Sources 140(1):1–11
15. Choi K, Peck D, Kim CS et al (2000) Water transport in polymer membranes for PEMFC. J Power Sources 86(1–2):197–201
16. Okada T, Kjelstrup RS, Moller HS et al (1996) Water and ion transport in the cation exchange membrane systems $NaCl-SrCl_2$ and $KCl-SrCl_2$. J Membr Sci 111(2):159
17. Fouquet N, Doulet C, Nouillant C et al (2006) Model based PEM fuel cell state-of-health monitoring via ac impedance measurements. J Power Sources 159(2):905–913
18. Agarwal P, Orazem ME, Garcia RLH (1995) Application of measurement models to impedance spectroscopy: III. Evaluation of consistency with the Kramers-Kronig relations. J Electrochem Soc 142(2):4159–4168
19. Andrzej L (2002) Modern aspects of electrochemistry: electrochemical impedance spectroscopy and its applications. Springer, Boston, 32: pp. 143–248
20. Brunetto C, Moschetto A, Tina G (2009) PEM fuel cell testing by electrochemical impedance spectroscopy. Electric Power Syst Res 79(1):17–26
21. Peter G, Frédéric J, Göran L et al (2003) Influence of the composition on the structure and electrochemical characteristics of the PEFC cathode. Electrochim Acta 48(28):4175–4187
22. Frédéric J, Göran L, Katarina W (2003) Transient techniques for investigating mass-transport limitations in gas diffusion electrodes: II. Experimental characterization of the PEFC cathode. J Electrochem Soc 150(12):A1711–A1717
23. Dale NV, Mann MD, Salehfar H et al (2010) AC impedance study of a proton exchange membrane fuel cell stack under various loading conditions. J Fuel Cell Sci Technol 7(3):0310101–03101010
24. Kitamura N, Manabe K, Nonobe Y et al, Development of water content control system for fuel cell hybrid vehicles based on AC impedance. SAE Technical Paper, 2010–01–1088
25. Springer TE, Zawodzinski TA, Gottesfeld S (1991) Polymer electrolyte fuel cell model. J Electrochem Soc 138(8):2334–2342
26. Springer TE, Wilson MS, Gottesfeld S (1993) Modeling and experimental diagnostics in polymer electrolyte fuel cells. J Electrochem Soc 140(2):3513–3526
27. Zawodzinski TA, Derouin C, Radzinski S et al (1993) Water uptake by and transport through Nafion® 117 membranes. J Electrochem Soc 140(4):1041–1047
28. Eikerling M, Kornyshev AA (1998) Modelling the performance of the cathode catalyst layer of polymer electrolyte fuel cells. J Electroanal Chem 453(1–2):89–106
29. Eikerling M, Kornyshev AA (1999) Electrochemical impedance of the cathode catalyst layer in polymer electrolyte fuel cells. J Electroanal Chem 475(2):107–123
30. Huang J, Zhang J (2016) Theory of impedance response of porous electrodes: simplifications, inhomogeneities, non-stationarities and applications. J Electrochem Soc 163(9):A1983–A2000
31. Michael BC (1975) An approximate model for mass transport with reaction in porous gas diffusion electrodes. Electrochim Acta 20(10):767–773
32. Raymond PI, Michael BC (1980) Voltage losses in fuel cell cathodes. J Electrochem Soc 127(7):1433–1440
33. Marr C, Li X (1999) Composition and performance modeling of catalyst layer in proton exchange membrane fuel cell. J Power Sources 77(1):17–27

34. Siegel JB, Stefanopoulou AG (2009) Through the membrane & along the channel flooding in PEMFCs. American control conference (ACC), pp 2666–2671
35. Hu J, Xu L, Li J et al (2016) Analytical calculation and evaluation of water transport through a proton exchange membrane fuel cell based on a one-dimensional model. Energy 111:869–883
36. Wang CY, Cheng P (1996) A multiphase mixture model for multiphase, multicomponent transport in capillary porous media—I. Model development. Int J Heat Mass Transp 39(17):3607–3618
37. Cheng P, Wang CY (1996) A multiphase mixture model for multiphase, multicomponent transport in capillary porous media—II. Numerical simulation of the transport of organic compounds in the subsurface. Int J Heat Mass Transp 39(17):3619–3632
38. Um S, Wang CY, Chen KS (2000) Computational fluid dynamics modeling of proton exchange membrane fuel cells. J Electrochem Soc 147(12):4485–4493
39. Um S, Wang CY (2000) Three dimensional analysis of transport and reaction in proton exchange membrane fuel cells. Amer Soc Mech Eng 366:19–22
40. Um S, Wang CY (2004) Three-dimensional analysis of transport and electrochemical reactions in polymer electrolyte fuel cells. J Power Sources 125(1):40–51
41. Meng H, Wang CY (2004) Electron transport in PEFCs. J Electrochem Soc 151(3):A358–A367
42. Meng H, Wang CY (2004) Large-scale simulation of polymer electrolyte fuel cells by parallel computing. Chem Eng Sci 59(16):3331–3343
43. Wang Y, Wang CY (2005) Transient analysis of polymer electrolyte fuel cells. Electrochim Acta 50(6):1307–1315
44. Wang Y, Wang CY (2006) Dynamics of polymer electrolyte fuel cells undergoing load changes. Electrochim Acta 51(19):3924–3933
45. Nguyen TV, Whit RE (1993) Water and heat management model for proton-exchange-membrane fuel cells. J Electrochem Soc 140(8):2178–2186
46. Yi JS, Nguyen TV (1998) Along-the-channel model for proton exchange membrane fuel cells. J Electrochem Soc 145(4):1149–1159
47. Yi JS, Nguyen TV (1999) Multicomponent transport in porous electrodes of proton exchange membrane fuel cells using the interdigitated gas distributors. J Electrochem Soc 146(1):38–45
48. Kazim A, Liu HT, Forges P (1999) Modelling of performance of PEM fuel cells with conventional and interdigitated flow fileds. J Appl Electrochem 29(12):1409–1416
49. Fuller TF, Newman J (1993) Water and thermal management in solid-polymer-electrolyte fuel cells. J Electrochem Soc 140(5):1218–1225
50. West AC, Fuller TF (1996) Influence of rib spacing in proton-exchange membrane electrode assemblies. J Appl Electrochem 26(6):557–565
51. Bao C, Ouyang M, Yi B (2006) Analysis of water management in proton exchange membrane fuel cells. Tsinghua Sci Technol 11(1):54–64
52. You L, Liu H (2002) A two-phase flow and transport model for the cathode of PEM fuel cells. Int J Hear Mass Transp 45(13):2277–2287
53. Wang Y, Basu S, Wang CY (2008) Modeling two-phase flow in PEM fuel cell channels. J Power Sources 179(2):603–617
54. Wang ZH, Wang CY, Chen KS (2001) Two-phase flow and transport in the air cathode of proton exchange membrane fuel cells. J Power Sources 94(1):40–50
55. Hu M, Zhu X, Wang M et al (2004) Three dimensional, two phase flow mathematical model for PEM fuel cell: Part II. Analysis and discussion of the internal transport mechanisms. Energy Conver Manag 45(11–12):1883–1916
56. Hu M, Gu A, Wang M et al (2004) Three dimensional, two phase flow mathematical model for PEM fuel cell: Part I. Model development. Energy Conver Manag 45(11–12):1861–1882
57. Maggio G, Recupero V, Pino L (2001) Modeling polymer electrolyte fuel cells: an innovative approach. J Power Sources 101(2):275–286
58. Squadrito G, Maggio G, Passalacqua E et al (1999) An empirical equation for polymer electrolyte fuel cell (PEFC) behavior. J Appl Electrochem 29(12):1449–1455

59. McKay D, Stefanopoulou AG (2004) Parametrization and validation of a lumped parameter diffusion model for fuel cell stack membrane humidity estimation. American control conference (ACC), pp. 816–821
60. Jiao K, Li X (2009) Three-dimensional multiphase modeling of cold start processes in polymer electrolyte membrane fuel cells. Electrochim Acta 54(27):6876–6891
61. Wang X, Tajiri K, Ahluwalia RK (2010) Water transport during startup and shutdown of polymer electrolyte fuel cell stacks. J Power Sources 195(19):6680–6687
62. Sinha PK, Wang CY (2008) Two-phase modeling of gas purge in a polymer electrolyte fuel cell. J Power Sources 183(2):609–618
63. Ding J, Mu Y, Zhai S et al (2016) Numerical study of gas purge in polymer electrolyte membrane fuel cell. Int J Heat Mass Transf 103:744–752
64. Meng H, A PEM fuel cell model for cold-start simulations. J Power Sources 178(1):141–150
65. Sundaresan M, Moore RM (2005) Polymer electrolyte fuel cell stack thermal model to evaluate sub-freezing startup. J Power Sources 145(2):534–545
66. Wakatake N, Tabe Y, Chikahisa T (2016) Water transport in ionomer and ice formation during cold startup with supercooled state in PEFC. ECS Trans 75(14):623–630
67. Lee CY, Lee YM, Lee SJ (2012) Local area water removal analysis of a proton exchange membrane fuel cell under gas purge conditions. Sensors 12(1):768–783
68. Sinha PK, Wang CY (2007) Gas purge in a polymer electrolyte fuel cell. J Electrochem Soc 154(11):B1158–B1166
69. Owejan PJ, Gagliardo JJ, Falta SR et al (2009) Accumulation and removal of liquid water in proton exchange membrane fuel cells. J Electrochem Soc 156(12):B1475–B1483
70. Robert SG (2000) Sensor needs and requirements for proton exchange membrane fuel cell systems and direct-injection engines. Livermore: Lawrence Livermore National Laboratory
71. Ous T, Arcoumanis C (2009) Visualisation of water accumulation in the flow channels of PEMFC under various operating conditions. J Power Sources 187(1):182–189
72. Liu X, Guo H, Ye F et al (2007) Water flooding and pressure drop characteristics in flow channels of proton exchange membrane fuel cells. Electrochim Acta 52(1):3607–3614
73. Liu X, Guo H, Ma C (2006) Water flooding and two-phase flow in cathode channels of proton exchange membrane fuel cells. J Power Sources 156(2):267–280
74. Yang X, Zhang F, Lubawy AL et al (2004) Visualization of liquid water transport in a PEFC. Electrochem Solid-State Lett 7(11):A408–A411
75. Zhang F, Yang X, Wang CY (2006) Liquid water removal from a polymer electrolyte fuel cell. J Electrochem Soc 153(2):A225–A232
76. Mosdale R, Gebel G, Pineri M (1996) Water profile determination in a running proton exchange membrane fuel cell using small-angle neutron scattering. J Membr Sci 118(2):269–277
77. Bellows RJ, Lin M, Arif M et al (1999) Neutron imaging technique for in situ measurement of water transport gradients within Nafion in polymer electrolyte fuel cells. J Electrochem Soc 146(3):1099–1103
78. Satija R, Jacobson DJ, Arif M et al (2004) In situ neutron imaging technique for evaluation of water management systems in operating fuel cells. J Power Sources 129(2):238–245
79. Karmer D, Zhang J, Shimoi R et al (2005) In situ diagnostic of two-phase flow phenomena in polymer electrolyte fuel cells by neutron imaging: Part A. Experimental, data treatment, and quantification. Electrochimica Acta 50(13):2603–2614
80. Kramer D, Lehmann E, Fei G et al (2005) An on-line study of fuel cell behavior by thermal neutrons. Nucl Instrum Methods Phys Res, Sect A 542(1–3):52–60
81. Pekula N, Heller K, Chuang P et al (2005) Study of water distribution and transport in a polymer electrolyte fuel cell using neutron imaging. Nucl Instrum Methods Phys Res Sect A 542(1–3):134–141
82. Chuang P, Turhan A, Heller AK et al (2005) The nature of flooding and drying in polymer electrolyte fuel cells. In ASME 2005 international conference on fuel cell science, engineering and technology. Amer Soc Mech Eng, 31–37

83. Ludlow DJ, Calebrese CM, Yu S et al (2006) PEM fuel cell membrane hydration measurement by neutron imaging. J Power Sources 162(1):271–278
84. Chen Y, Peng H, Hussey DS et al (2007) Water distribution measurement for a PEMFC through neutron radiography. J Power Sources 170(2):376–386
85. Siegel JB, McKay AD, Stefanopoulou AG (2008) Modeling and validation of fuel cell water dynamics using neutron imaging. American control conference (ACC), pp. 2573–2578
86. Siegel JB, Stefanopoulou AG, Yesilyurt S (2009) Extracting model parameters and paradigms from neutron imaging of dead-ended anode operation. Proceedings of the 7th international conference on fuel cell science, engineering and technology. American Society of Mechanical Engineers, pp 439–446
87. Murakawa H, Sugimoto K, Kitamura N et al (2015) Visualization of water accumulation process in polymer electrolyte fuel cell using neutron radiography. Phys Procedia 69:607–611
88. Salva JA, Iranzo A, Rosa F et al (2016) Validation of cell voltage and water content in a PEM (polymer electrolyte membrane) fuel cell model using neutron imaging for different operating conditions. Energy 101:100–112
89. Steiner NY, Hissel D, Candusso D et al (2011) Diagnosis of polymer electrolyte fuel cells failure modes (flooding & drying out) by neural networks modeling. Int J Hydrogen Energy 36(4):3067–3075
90. Ito K, Ashikaga K, Masuda H et al (2008) Estimation of flooding in PEMFC gas diffusion layer by differential pressure measurement. J Power Sources 175(2):732–738
91. He W, Lin G, Nguyen TV (2003) Diagnostic tool to detect electrode flooding in Proton-Exchange-Membrane Fuel Cells. AIChE J 49(12):3221–3228
92. Nguyen TV, Liu G, Ohn N et al (2008) Measurement of capillary pressure property og gas diffusion media used in proton exchange membrane fuel cells. Electrochem Solid-State Lett 11(8):B127–B121
93. He W, Yi JS, Nguyen TV (2000) Two-phase flow model of the cathode of PEM fuel cells using interdigitated flow fields. AIChE J 46(10):2053–2064
94. Knobbe MW, He W, Chong P et al (2004) Active gas management for PEM fuel cell stacks. J Power Sources 138(1–2):94–100
95. Song M, Pei P, Zha H et al (2014) Water management of proton exchange membrane fuel cell based on control of hydrogen pressure drop. J Power Sources 267(1):655–665
96. Pei P, Ouyang M, Lu Q et al (2004) Testing of an automotive fuel cell system. Int J Hydrogen Energy 29(10):1001–1007
97. Lee D, Bae J (2009) Evaluation of the net water transport through electrolytes in proton exchange membrane fuel cell. J Power Sources 191(2):390–399
98. Janssen GJM, Overvelde MLJ (2000) Water transport in the proton-exchange-membrane fuel cell: measurements of the effective drag coefficient. J Power Sources 101(1):117–125
99. Li Y, Pei Y, WY et al (2018) Approaches to avoid flooding in association with pressure drop in proton exchange membrane fuel cells. Appl Energy, 224:42–51
100. Haluk G, Arcak M, Barbir F (2006) An algorithm for estimation of membrane water content in PEM fuel cells. J Power Sources 157(1):389–394
101. Arcak M, Haluk G, Pedersen LM et al (2004) A nonlinear observer design for fuel cell hydrogen estimation. IEEE Trans Control Syst Technol 12(1):101–110
102. Parthasarathy A, Dave B, Srinivasan S et al (1992) The platinum microelectrode/Nafion interface: an electrochemical impedance spectroscopic analysis of oxygen reduction kinetics and Nafion characteristics. J Electrochem Soc 139(6):1634–1641
103. Wu J, Yuan X, Wang H et al (2008) Diagnostic tools in PEM fuel cell research: Part I Electrochemical techniques. Int J Hydrogen Energy 33(6):1735–1746
104. Yuan X, Wang H, Sun JC et al (2007) AC impedance technique in PEM fuel cell diagnosis—a review. Int J Hydrogen Energy 32(17):4365–4380
105. Springer TE, Zawodzinski TA, Wilson MS et al (1996) Characterization of polymer electrolyte fuel cells using AC impedance spectroscopy. J Electrochem Soc 143(2):587–599
106. Yuan X, Sun JC, Blanco M et al (2006) AC impedance diagnosis of a 500 W PEM fuel cell stack: Part I: Stack impedance. J Power Sources 161(2):920–928

107. Yuan X, Sun JC, Wang H et al (2006) AC impedance diagnosis of a 500 W PEM fuel cell stack: Part II: Individual cell impedance. J Power Sources 161(2):929–937
108. Yan X, Hou M, Sun L et al (2007) AC impedance characteristics of a 2kW PEM fuel cell stack under different operating conditions and load changes. Int J Hydrogen Energy 32(17):4358–4364
109. Aaron D, Yiacoumi S, Tsouris C (2008) Effects of proton-exchange membrane fuel-cell operating conditions on charge transfer resistances measured by electrochemical impedance spectroscopy. Sep Sci Technol 43(9–10):2307–2320
110. Malevich D, Halliop E, Peppley BA et al (2008) Effect of relative humidity on electrochemically active area and impedance response of PEM fuel cell. ECS Trans 16(2):1763–1774
111. Malevich D, Halliop E, Saha MS et al (2009) Effect of microporous layer on impedance response of PEM fuel cell fed with H2 and N2. Meeting abstracts. ECS Transactions, 977
112. Mathias MF, Grot SA (2002) System and method for controlling the humidity level of a fuel cell: U.S.A., 6376111B1. 2002–04–23
113. Cha D, Ahn JH, Kim HS et al (2015) Effects of clamping force on the water transport and performance of a PEM (proton electrolyte membrane) fuel cell with relative humidity and current density. Energy 93(2):1338–1344
114. Asghari S, Mokmeli A, Samavati M (2010) Study of PEM fuel cell performance by electrochemical impedance spectroscopy. Int J Hydrogen Energy 35(17):9283–9290
115. Qi Z, Kaufman A (2002) Activation of low temperature PEM fuel cells. J Power Sources 111(1):181–184
116. Xu Z, Qi Z, He C et al (2006) Combined activation methods for proton-exchange membrane fuel cells. J Power Sources 156(2):315–320
117. Freire TJP, Gonzalez ER (2001) Effect of membrane characteristics and humidification conditions on the impedance response of polymer electrolyte fuel cells. J Electroanal Chem 503(1–2):57–68
118. Oszcipok M, Riemann D, Kronenwett U et al (2005) Statistic analysis of operational influence on the cold start behavior of PEM fuel cells. J Power Sources 145(2):407–415
119. Randles JEB (1947) Kinetics of rapid electrode reactions. Faraday Soc 1:11–19
120. Randles JEB, Somerton KW (1952) Kinetics of rapid electrode reactions. Faraday Soc 48:937–950
121. Grahame DC (1952) Mathematical theory of the Faradaic admittance. J Electrochem Soc 99:C370–C385
122. Cimenti M, Bessarabov D, Tam M et al (2010) Investigation of proton transport in the catalyst layer of PEM fuel cells by electrochemical impedance spectroscopy. ECS Trans 28(23):147–157
123. Malevich D, Pharoah J, Peppley B et al (2011) On the determination of PEM fuel cell catalyst layer resistance from impedance measurement in H2/N2 cells. ECS Trans 41(1):721–732
124. Jang JH, Jeon S, Cho JH et al (2009) Complex capacitance analysis of ionic resistance and interfacial capacitance in PEMFC and DMFC catalyst layers. J Electrochem Soc 156(11):B1293–B1300
125. Young AP, Stumper J, Gyenge E (2009) Characterizing the structural degradation in a PEMFC cathode catalyst layer: carbon corrosion. J Electrochem Soc 156(8):B913–B922
126. Liu Y, Murphy MW, Baker DR et al (2009) Proton conduction and oxygen reduction kinetics in PEM fuel cell cathodes: effects of ionomer-to-carbon ratio and relative humidity. J Electrochem Soc 156(8):B970–B980
127. Alkire RC, Bartlett PN, Lipkowski J (2015) Electrochemistry of carbon electrodes. Wiley-VCH, Weinheim
128. Makharia R, Mathias MF, Baker DR (2005) Measurement of catalyst layer electrolyte resistance in PEFCs using electrochemical impedance spectroscopy. J Electrochem Soc 152(5):A970–A977
129. Levillain E, Demortier A, Lelieur JP (1995) Electrochemical impedance of solutions of polysulfides in liquid ammonia: experimental evidence for the Gerischer impedance. J Electroanal Chem 394(1–2):103–115

130. Dhanda A, Pitsch H, O'Hayre D (2011) Diffusion impedance element model for the triple phase boundary. J Electrochem Soc 158(8):B877–B884
131. Boukamp B, Bouwmeester HJM (2003) Interpretation of the Gerischer impedance in solid state ionics. Solid State Ionics 157(1–4):29–33
132. Danzer MA, Hofer EP (2008) Electrochemical parameter identification-An efficient method for fuel cell impedance characterization. J Power Sources 183(1):55–61
133. Andreasen SJ, Vang JR, Kaer SK (2011) High temperature PEM fuel cell performance characterisation with CO and CO_2 using electrochemical impedance spectroscopy. Int J Hydrogen Energy 36(16):9815–9830
134. Danzer MA, Hofer EP (2009) Analysis of the electrochemical behaviour of polymer electrolyte fuel cells using simple impedance models. J Power Sources 190(1):25–33
135. Cimenti M, Tam M, Stumper J (2009) High frequency artifacts in electrochemical impedance spectroscopy measurements on PEM fuel cells. Electrochem Solid-State Lett 12(9):B131–B134
136. Schneider IA, Freunberger SA, Kramer D et al (2007) Oscillations in gas channels: Part I. The forgotten player in impedance spectroscopy in PEFCs. J Electrochem Soc 154(4):B383-B388
137. Mikhailenko SD, Guiver MD, Kaliaguine S (2008) Measurements of PEM conductivity by impedance spectroscopy. Solid State Ionics 179(17–18):619–624
138. Lee CH, Park HB, Lee YM et al (2005) Importance of proton conductivity measurement in polymer electrolyte membrane for fuel cell application. Ind Eng Chem Res 44:7617–7626
139. Bender G, Wilson MS, Zawodzinski TA (2003) Further refinements in the segmented cell approach to diagnosing performance in polymer electrolyte fuel cells. J Power Sources 123(2):163–171
140. Schneider IA, Kuhn H, Wokaun A et al (2005) Study of water balance in a polymer electrolyte fuel cell by locally resolved impedance spectroscopy. J Electrochem Soc 152(12):A2383–A2389
141. Schneider IA, Kramer D, Wokaun A et al (2005) Spatially resolved characterization of PEFCs using simultaneously neutron radiography and locally resolved impedance spectroscopy. Electrochem Commun 12(7):1393–1397
142. Brunetto C, Tina G, Squadrito G et al (2004) PEMFC diagnostics and modeling by electrochemical impedance spectroscopy. Proceedings of the 12th IEEE Mediterranean Electrotechnical conference. IEEE Electrotechnical conference, 3:1045–1050
143. Brunetto C, Moschettob A, Tina G (2009) PEM fuel cell testing by electrochemical impedance spectroscopy. Electric Power Syst Res 79:17–26
144. Wasterlain S, Harel F, Candusso D et al (2009) First results obtained with an impedance meter developed for the diagnosis of large proton exchange membrane fuel cell stack. Advanced electromechanical motion systems & electric drives joint symposium. Electromotion, pp 1–6
145. Wasterlain S, Candusso D, Harel F et al (2010) Diagnosis of a fuel cell stack using electrochemical impedance spectroscopy and Bayesian Networks. IEEE 2010 vehicle power and propulsion conference (VPPC). IEEE Computer Society, pp 1–6
146. Wasterlain S, Candusso D, Harela F et al (2011) Development of new test instruments and protocols for the diagnostic of fuel cell stacks. J Power Sources 196(12):5325–5333
147. Petrone R (2014) Electrochemical impedance spectroscopy for the on-board diagnosis of PEMFC via on-line identification of equivalent circuit model parameters. University of Salerno, Italy
148. Petrone R, Zheng Z, Hissel D et al (2013) Implementation of EIS measurements on an embedded commercial system. Poster presentation 5th international conference FDFC
149. Gasda MD, Misiewicz M, Prescott G et al (2006) Technique and apparatus to measure a fuel cell parameter: U.S.A., 7099787B2. 2006–08–29
150. Erich R (2006) Method for monitoring the operating state of a fuel cell stack: U.S.A. 20060078788A1. 2006–04–13
151. Peter P, Kartharina R (2015) Method for determining critical operating states in a fuel cell stack. U.S.A., 20150153418A1. 2015–06–04

152. Ramschak E, Fouquet N, Brandstaetter H et al (2010) Online FC monitoring and identification of operating conditions including lifetime measurement. Sustainable mobility revolution: 25th world battery, hybrid and fuel cell electric vehicle symposium and exhibition. Electric Vehicle Systems
153. Narjiss A, Depernet D, Candusso D et al (2008) Online diagnosis of PEM fuel cell. Power electronics and motion control conference 3rd. EPE-PEMC, pp 734–739
154. Narjiss A, Depernet D, Gustin F et al (2008) High frequency power converter for fuel cell stacks parallel association. IEEE 2008 vehicle power and propulsion conference (VPPC). IEEE Computer Society, pp 1-4
155. Hinaje M, Sadli I, Martin JP et al (2009) Online humidification diagnosis of a PEMFC using a static DC-DC converter. Int J Hydrogen Energy 13(6):2718–2723
156. Dotelli G, Ferrero R, Stampino PG et al (2013) Inverter ripple as a diagnostic tool for ohmic resistance measurements on PEM fuel cells. IEEE 2013 applied measurements for power systems (AMPS). IEEE Computer Society, pp 156–161
157. Dotelli G, Ferrero R, Stampino PG et al (2014) Diagnosis of PEM fuel cell drying and flooding based on power converter ripple. IEEE Trans Instrum Meas 63(10):2341–2348
158. Dotelli G, Ferrero R, Stampino PG et al (2015) Low-cost PEM fuel cell diagnosis based on power converter ripple with hysteresis control. IEEE Trans Instrum Meas 99:1–8
159. Dotelli G, Ferrero R, Stampino PG et al (2015) PEM fuel cell drying and flooding diagnosis with signals injected by a power converter. IEEE Trans Instrum Meas 64(8):2064–2071
160. OH D, Sun H (2010) Method and apparatus for diagnosing deterioration of fuel cell: U.S.A., 20100286939A1. 2010–11–11
161. Koichi K, Sekine S (2011) Development of fuel cell hybrid vehicle in TOYOTA[R]. SAE Technical Paper, 2011–39–7238
162. Kota M, Takeshi M, Takahiko H (2010) Fuel cell system: U.S.A., 20100013490A1. 2010–01–21
163. Kota M, Mashiro S (2011) Fuel cell system: U.S.A., 20110300461A1. 2011–12–08
164. Kota M, Masahiro H (2007) Fuel cell system and vehicle: Japan, 2007012414A. 2007–01–18
165. Kota M, Hikaru A (2013) Impedance measurement device: WIPO, PCT2013042200. 2013–03–28
166. Hasuka Y, Sekine H, Katano K et al (2015) Development of Boost converter for MIRAI[R]. SAE Technical Paper, 2015–01–1170
167. Takahiko H, Kota M (2012) Chopper circuit, DC/DC converter and fuel cell system: U.S.A., 2012326687A1. 2012–12–27
168. Kim JH (2015) Apparatus and method for measuring internal ohmic resistance of fuel cell system: U.S.A., 20150050524A1. 2015–02–19
169. Sakai M, Kanagawa A (2015) Device for measuring impedance of laminated battery: Europe, 2908149 B1. 2015–08–19
170. Weng F, Su A, Hsu C (2007) The study of the effect of gas stoichiometric flow rate on the channel flooding and performance in a transparent fuel cell. Int J Hydrogen Energy 32(6):666–676
171. Andrew DB, Matthew HF (2000) Fuel cell flooding detection and correction: U.S.A., 6103409. 2000–08–15
172. Paul R, Felix B, Chris O et al (2004) Operational aspects of a large PEFC stack under practical conditions. J Power Sources 128(2):208–217
173. Barbir F, Husar A, Venkataraman R (2001) Pressure drop as a diagnostic tool for PEM fuel cells. Meeting Abstracts. Electrochemical Society
174. Barbir F, Haluk G, Wang X (2005) Relationship between pressure drop and cell resistance as a diagnostic tool for PEM fuel cells. J Power Sources 141(1):96–101
175. Kurz T, Hakenjos A, Krämer J et al (2008) An impedance-based predictive control strategy for the state-of-health of PEM fuel cell stacks. J Power Sources 180(2):742–747
176. Karnik AY, Stefanopoulou AG, Sun J (2007) Water equilibira and management using a two-volume model of a polymer electrolyte fuel cell. J Power Sources 164(2):590–605

177. Fang C, Li J, Xu L et al (2015) Model-based fuel pressure regulation algorithm for a hydrogen-injected PEM fuel cell engine. Int J Hydrogen Energy 40(39):13566–13575
178. Li J, Xu L, Cheng S et al (2012) The fuel cell system with recirculation of waste gas of cathode and anode: China, 201210586133.2
179. Kim BJ, Kim MS (2012) Studies on the cathode humidification by exhaust gas recirculation for PEM fuel cell. Int J Hydrogen Energy 37(5):4290–4299
180. Hu J, Xu L, Li J et al (2014) Water management in a self-humidifying PEM fuel cell system by exhaust gas recirculation. Transportation electrification Asia-Pacific (ITEC Asia-Pacific), 2014 IEEE conference and expo. IEEE, 2014, 1: 1–6
181. Zhao X, Xu L, Fang C et al (2018) Study on voltage clamping and self-humidification effects of pem fuel cell system with dual recirculation based on orthogonal test method. Int J Hydrogen Energy 43(33):16268–16278
182. Bao C, Ouyang M, Yi B (2006) Modeling and optimization of the air system in polymer exchange membrane fuel cell systems. J Power Sources 156(2):232–243
183. Hasegawa T, Imanishi H, Nada M et al (2016) Development of the Fuel Cell System in the Mirai FCV. SAE Technical Paper, 2016–01–1185
184. Chen J, Siegel JB, Stefanopoulou AG et al (2013) Optimization of purge cycle for dead-ended anode fuel cell operation. Int J Hydrogen Energy 38(12):5092–5105
185. Marsuura T, Siegel JB, Stefanopoulou AG (2012) Experimental investigation of degradation in PEMFC with dead-ended anode operation. Meeting Abstracts. The Electrochemical Society 6:315–315
186. Yesilyurt S, Siegel J B, Stefanopoulou AG (2012) Modeling and experiments of voltage transients of polymer electrolyte membrane fuel cells with the dead-ended anode. J Fuel Cell Sci Technol 9(2):021012–1–7
187. Chen J, Siegel JB, Stefanopoulou AG (2011) Nitrogen blanketing front equilibria in dead end anode fuel cell operation. American control conference (ACC), pp 1524–1529
188. Chen J, Siegel JB, Matsuura T et al (2011) Carbon corrosion in PEM fuel cell dead-ended anode operations. J Electrochem Soc 158(9):B1164–B1174

Chapter 2
Modeling of Water Content in MEA for the PEM Fuel Cell

2.1 Introduction

In a PEM fuel cell, the MEA lies in the center and is sandwiched between the cathode gas diffusion layer (CGDL) and the anode gas diffusion layer (AGDL). The electrochemical reaction happens both in the cathode catalyst layer (CCL) formed on one side of the MEA and in the anode catalyst layer (ACL) formed on the other side of the MEA. The water content in the MEA dominates the conductivity of the proton, particularly in the polymer electrolyte solution, and its distribution through MEA is affected by the electro-osmotic drag effect and the back diffusion due to the gradient of local water content.

It is known that when liquid water in the MEA is nearly saturated, the conductivity of proton is high enough but the accumulated water in the cathode gas diffusion layer will block the oxygen transport, thus lowering the local oxygen concentration in the cathode catalyst layer. On the contrary, when the liquid water in the MEA is very low, the conductivity of proton is significantly increased but the resistance of oxygen transport in the cathode gas diffusion layer is largely decreased, thus improving the local oxygen concentration in the cathode catalyst layer. The local oxygen concentration determines the activation overvoltage in either case. Through these two common phenomena, it's clear that the characteristic of the MEA reflects the variation of water content in a fuel cell directly and the electrochemical impedance spectroscopy as a tool will facilitate the research of mechanism of the MEA.

In this chapter, the model of water content in MEA will be established, to quantify the influence of water content on the MEA characteristic and its impedance spectroscopy. This chapter is organized as follows: in Sect. 2.2, the mathematical connection between the polarization curve of a fuel cell and the equivalent circuit model is analyzed, to quantify the voltage loss of the fuel cell in terms of the impedance spectroscopy; in Sect. 2.3, a simplified mode of the MEA is presented to study the internal recirculation of water in a fuel cell based on a special structure as well as

P. Hong, *Water Content Estimation and Control of PEM Fuel Cell Stack and the Individual Cell in Vehicle*, Springer Theses, https://doi.org/10.1007/978-981-16-8814-0_2

the performance of this cell; in Sect. 2.4, the model of water content in MEA is developed and some interesting result is found like the evolution of the equivalent circuit model with the water content; in Sect. 2.5, our research is summarized.

2.2 Connection Between Polarization Curve and Equivalent Circuit Model

The principal of a single-flow-channel fuel cell is presented (Fig. 2.1) on basis of the Fig. 1.4. The flow of oxygen in the cathode flow channel and that of hydrogen in the anode flow channel are designed in counter-flow mode. Under predefined condition of fuel supply and coolant temperature, the output voltage of the fuel cell varies with the current density and the diagram describing their relation is called the polarization curve (Fig. 2.2).

The polarization curve is usually described, as Eq. 2.1, and the output voltage V of a cell is function of the open circuit voltage V_{OC} (Eq. 2.2), the activation overvoltage η_{act} (Eq. 2.3), the ohm overvoltage η_{ohm} (Eq. 2.4) and the concentration overvoltage η_{con} (Eq. 2.5). Whereas A and B are fitted coefficient, E_0 the theoretical Nernst voltage, i the current density, i_{cross} the crossover current density, i_0 the exchange current density, R_{ohm} the ohm resistance, i_{Limit} the limit current density, R the gas constant, F the Faradic constant, T the operating temperature, P_{Oxy}, P_{Hy} and P_W the theoretical partial pressure of oxygen, hydrogen and water vapor in the catalyst layer correspondingly. Please notice that the water is assumed to be in the vapor state. The limit current density i_{Limit} is determined by the oxygen concentration $C_{Oxy,1}$ in the cathode gas flow channel and the oxygen concentration $C_{Oxy,6}$ in the cathode catalyst layer is zero.

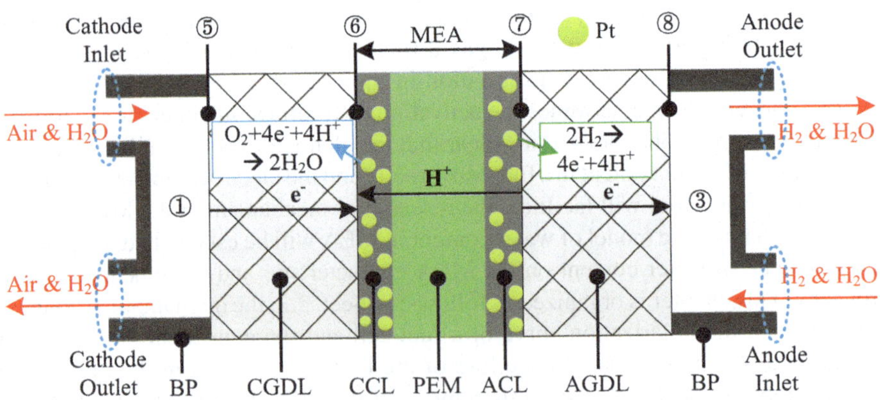

Fig. 2.1 Principal of a single-flow-channel fuel cell

Fig. 2.2 Polarization curve of a fuel cell

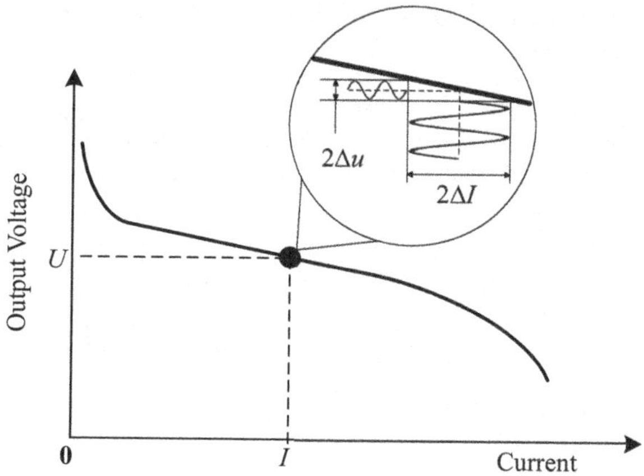

$$V = V_{OC} - \eta_{act} - \eta_{ohm} - \eta_{con} \tag{2.1}$$

$$V_{OC} = E_0 + \frac{RT}{2F} \ln \frac{P_{Oxy}^{1/2} P_{Hy}}{P_W} \tag{2.2}$$

$$\eta_{act} = A \ln \frac{i + i_{cross}}{i_0} \tag{2.3}$$

$$\eta_{ohm} = i R_{Ohm} \tag{2.4}$$

$$\eta_{con} = -B \ln \frac{i_{Limit} - i}{i_{Limit}} \tag{2.5}$$

$$N_{Oxy} = C_{Oxy,1} - C_{Oxy,6} = \frac{P_{Oxy,1} - P_{Oxy,6}}{RT} = \frac{i}{4F} \tag{2.6}$$

The oxygen consumption rate N_{Oxy} is proportional to current density of the cell as Eq. 2.6, whereas the $P_{Oxy,1}$ is the oxygen partial pressure in the cathode gas flow channel, the $P_{Oxy,6}$ the oxygen partial pressure in the cathode catalyst layer and the $D_{Oxy,eff}$ the effective diffusion coefficient of oxygen in the cathode gas diffusion layer. The activation overvoltage and the concentration overvoltage can be re-organized as Eq. 2.7. It can be found that the oxygen concentration $C_{Oxy,1}$ increases as the oxygen concentration $C_{Oxy,6}$ increases with the current density fixed, but the activation and concentration overvoltage will be decreased. Besides, the activation and concentration overvoltage increases as the current density i increases with the oxygen concentration $C_{Oxy,1}$ fixed. In hence, the concentration overvoltage is used to describe the overvoltage loss resulting from the difference between the oxygen concentration in cathode gas flow channel and that in cathode catalyst layer, but in fact it is still a part of the activation overvoltage in cathode catalyst layer. In following chapters, no differentiation will be made between them and they are calculated as the activation overvoltage.

$$\eta_{act} + \eta_{con} = A \ln\left(\frac{i}{i_0}\left(\frac{C_{Oxy,1}}{C_{Oxy,6}}\right)^{\beta}\right) \tag{2.7}$$

$$\begin{aligned} j_{An} = j_{0,An,ref}\left(\frac{c_{Hy}}{c_{Hy,ref}}\right)^{0.5} \times \\ \left[\exp\left(\frac{2\alpha_{An}F}{RT}\eta_{act,An}\right) - \exp\left(-\frac{2\alpha_{Ca}F}{RT}\eta_{act,An}\right)\right] \end{aligned} \tag{2.8}$$

$$\begin{aligned} j_{Ca} = j_{0,Ca,ref}\frac{c_{Oxy,6}}{c_{Oxy,ref}} \times \\ \left[-\exp\left(\frac{4\alpha_{An}F}{RT}\eta_{act,Ca}\right) + \exp\left(-\frac{4\alpha_{Ca}F}{RT}\eta_{act,Ca}\right)\right] \end{aligned} \tag{2.9}$$

$$\begin{cases} A = B = \frac{RT}{4\alpha_{Ca}F} \\ i_0 = j_{0,Ca,ref}\frac{c_{Oxy,1}}{c_{Oxy,ref}} \end{cases} \tag{2.10}$$

In the cathode catalyst layer, the Butler-Volmer equation is used to quantify the activation overvoltage, as Eq. 2.8, and it is the same for the anode catalyst layer, as Eq. 2.9. The activation overvoltage of hydrogen oxidation reaction is far lower than that of oxygen reduction reaction such that the voltage loss in the anode catalyst layer is usually neglected and this will be applied to research in following chapters.

Based on Eqs. 2.7 and 2.9, part of terms in the polarization curve satisfies the Eq. 2.10 and the term $C_{Oxy,6}$ in Eq. 2.7 can be neglected. In part of the MEA where no electrochemical reaction happens, the voltage loss η_{ohm} only results from the conduction of proton and electronic and it is product of the current density and area-related resistance.

$$i = i_{stack} + \Delta I \sin(\omega t + \beta_1) \tag{2.11}$$

$$V = V_{stack} + \Delta V \sin(\omega t + \beta_2) \tag{2.12}$$

When the fuel cell operates in steady state, a small current excitation with amplitude ΔI is superimposed on the steady output current i and it is a sinusoidal signal, as Eq. 2.11, whereas ω is the angular frequency, t time and θ_1 the initial phase. Accordingly, the output voltage V of the cell is as Eq. 2.12 based on the linearity response of the fuel cell to small stimulation, whereas ΔV is amplitude of the dynamic voltage response and θ_2 the initial phase.

$$\left(\frac{d\eta_{act}}{di}\right)_{\omega \to 0} = R_F \tag{2.13}$$

$$Z = Laplace\left[\frac{dV}{di}\right] = \frac{1}{R_F C\omega j + 1} + R_{Ohm} \tag{2.14}$$

$$\left(\frac{dV}{di}\right)_{\omega\to\infty} = R_{Ohm} \tag{2.15}$$

$$\left(\frac{dV}{di}\right)_{\omega\to 0} = R_F + R_{Ohm} \tag{2.16}$$

The amplitude ΔV and the initial phase θ_2 of the dynamic voltage response will change with the variation of the angular frequency ω, which results from the double-layer capacitor and the Faradic resistance in the catalyst layer where the electrochemical reaction happens. To have a direct view of the charge–discharge phenomenon and the steady-state voltage loss of the fuel cell, Randles put forward a circuit where the Faradic resistance is in parallel with the double-layer capacitor (Fig. 2.3a). The Faradic resistance is calculated as Eq. 2.13 and the equivalent resistance of the fuel cell derived as Eq. 2.14. If the angular frequency increases to the positive infinity, the equivalent resistance is as Eq. 2.15 and it's called the high frequency resistance (HFR). If the angular frequency decreases to zero, the equivalent resistance is presented as Eq. 2.16 and this is the low frequency resistance (LFR). Both phases of the HFR and the LFR are zero. According to Eq. 2.16 and Fig. 2.3a, we can find that the LFR is equal to the absolute value of the slope of the polarization curve.

The equivalent circuit model (Fig. 2.3a) provides direct insight into the voltage loss in the MEA. However, for a fuel cell in application, the contact resistance between the bipolar plate and the CGDL, that between the CGDL and the CCL, that between the bipolar plate and the AGDL and that between the AGDL and ACL need to be considered. On this basis, the equivalent circuit model is improved (Fig. 2.3b). The equivalent circuit model is further optimized as a result of the nonuniformity between cells (Fig. 2.3c).

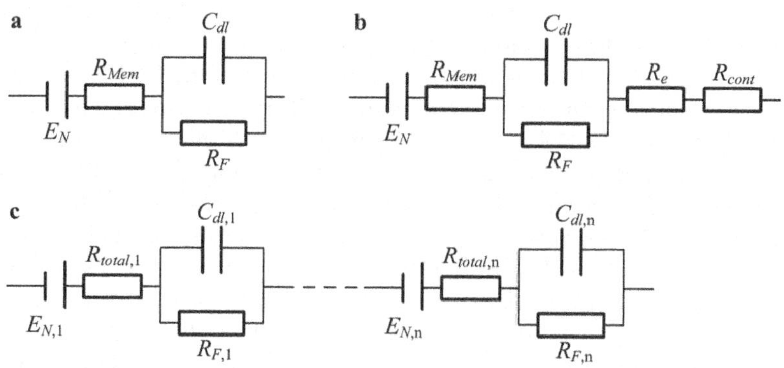

Fig. 2.3 Equivalent circuit model of the fuel cell: **a** Simplified model, **b** Improved model, **c** The stack model

The equivalent circuit model quantifies dynamic response and steady state characteristic of the MEA. To be specific, due to the relatively slow dynamic of mass transport, the concentration of oxygen and water in the electrochemical reaction zone cannot follow the current density without delay when a current excitation at medium or higher frequency is superimposed on a fuel cell. Here, the medium or higher frequency is compared to that of the mass transport which is inversely proportional to its time constant. In this case, the average concentration of oxygen and water keeps constant in just one period of the current excitation and the oxygen consumption and water generation keeps at a fixed rate proportional to the current density in accordance with average of integration of Eq. 2.11 during this time. The conclusion is that the mass transport has no effect on the current resistance at medium or higher frequency.

On the other hand, the concentration of oxygen and water in the electrochemical zone can follow the current density with frequency-related delay when a current excitation at low frequency is superimposed on a fuel cell. Due to the relatively fast dynamic of the double-layer capacitor and the Faradic resistance, the electrochemical reaction tracks the concentration of oxygen and water. According to Eq. 2.7, the voltage loss varies with the concentration of oxygen and water and the lower the excitation frequency, the larger impact of mass transport on the resistance.

Based on Fig. 2.3b and Eq. 2.15, the HFR describes the voltage loss related to conduction of proton and electronic. The fact is that the resistance of electronic conduction is closely related to clamping pressure over the fuel cell and this factor changes slowly and limitedly. That is to say, the effect of the operation condition and water content of the fuel cell on the electronic resistance is small. On the contrary, the proton conductivity in the MEA is closely related to the operating condition and water content of the fuel cell and in practical application, the resistance of proton conduction varies significantly with them. Therefore, for a fuel cell in steady operation, the variation of proton conduction dominates that of the HFR and this is the key to studying water content of a fuel cell.

Based on Eqs. 2.7 and 2.13, the Faradic resistance R_F reflects oxygen concentration at the electrochemical reaction site with the operating condition and the current density constant. In the situation where liquid water accumulates in the fuel cell due to poor water drainage, the effective diffusion area of the CGDL is decreased. Consequently, the oxygen concentration in the CCL is decreased and the activation overvoltage increases as well as the Faradic resistance R_F and vice versa.

Definitely, the polarization curve and the equivalent circuit model is mathematically equal to each other and both can describe characteristic of the fuel cell. The HFR and LFR are key parameters to studying water content of the fuel cell. Furthermore, the voltage loss of the electrochemical reaction and the proton conduction happens mainly in the MEA and the electrochemical impedance spectroscopy shows the electrical characteristic of the MEA in a fuel cell, so the MEA is the linkage between water content and the fuel cell characteristic. Specifically, the Faradic resistance R_F is the activation overvoltage in MEA divided by current when analyzing the LFR. The concentration of oxygen and water in the MEA cannot follow the current excitation when analyzing the HFR due to their slow dynamic, such that the oxygen concentration and oxygen and water in the CGDL can be considered as constant and this will contribute to setting up a model for the MEA in terms of boundary conditions.

2.3 Internal Recirculation of Water Content in a Fuel Cell

The fuel cell for the vehicle power source is usually of large surface area and effective area for large current. The distribution of current density is not uniform along flow channels in the MEA as well as that of oxygen, nitrogen and water in the CGDL and that of local temperature in the fuel cell. It is very necessary to choose an appropriate operating condition while designing the structure to improve uniformity and durability. In the fuel cell system, the humidifier is one of key accessories installed to increase relative humidity of the air and hydrogen supplied to the fuel cell stack. It is certain that the humidifier will improve performance of the fuel cell in most cases but when considering system volume and cost, the humidifier is not always on the check list. For a system without a humidifier, the air entering the cathode inlet of the fuel cell stack is the same as that in the environment and through an intercooling device, the air temperature is around the target operating temperature of the cell.

Table 2.1 lists the saturation pressure of water vapor at different temperature. For example, the environment temperature is 20 °C and the relative humidity is 100%. If the fuel cell stack operates at 60 °C and the absolute pressure of air at the cathode inlet is 150 kPa, the relative humidity at the cathode inlet is just 17.4%. A higher operating temperature and a lower temperature and relative humidity of the environment will lead to the lower relative humidity of air at the cathode inlet. In such a fuel cell, the reaction area close to the cathode inlet tends to be dried out while the reaction area close to the cathode outlet tends to be flooded. Special structure design has been tried to relieve the problem in this situation by Toyota Motor Corporation [1, 2] and it is the research object in this section (Fig. 2.4).

Table 2.1 Saturation pressure of water vapor

Temperature (°C)	Saturation pressure (kPa)	Temperature (°C)	Saturation pressure (kPa)
0	0.6113	40	7.3814
5	0.8726	45	9.5895
10	1.2281	50	12.3440
15	1.7956	55	15.6520
20	2.3388	60	19.9320
25	3.1690	65	25.0220
30	4.2455	70	31.1760
35	5.6267	75	38.5630

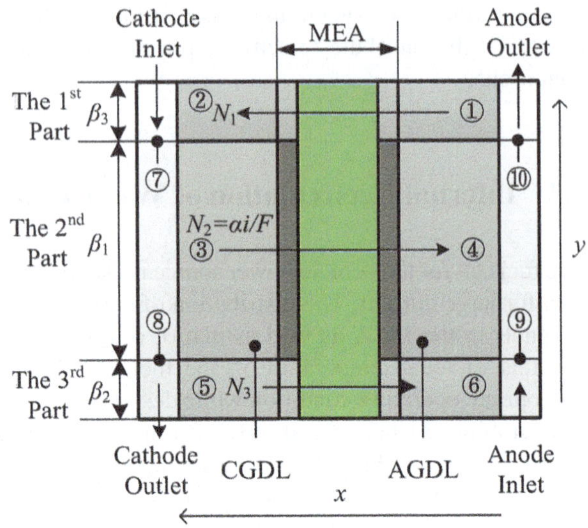

Fig. 2.4 Principle of internal water recirculation of the fuel cell without external humidifier

In this fuel cell, the zone ② close to the cathode inlet is designed to promote the evaporation of water while the zone ⑤ close to the cathode outlet is enhanced to be more hydrophilic to absorb water from mixed air and water. Correspondingly, the zone ① close to the hydrogen outlet is more hydrophilic to absorb water from the mixed hydrogen and water while the zone is⑥ close to the hydrogen inlet is designed to promote the evaporation of water.

The effective operating area of the fuel cell is assumed to be the zone ③ and ④ and the MEA in between to simplify the model. When the fuel cell runs, the water recirculation is formed in such a way that the air flow carries water from the cathode inlet to the outlet, the water diffuses to the anode inlet from the cathode outlet driven by the gradient of water concentration through the MEA, the hydrogen flow carries water from the anode inlet to the outlet, and finally the water diffuses to the cathode inlet from the anode outlet driven by the gradient of water concentration through the MEA.

The effective area S of the fuel cell includes the zone ① with area $\beta_3 S$, the zone ④ with area $\beta_1 S$ and the zone ⑥ with area $\beta_2 S$. The zone ②, ③ and ⑤ has the same area as the zone①, ④ and ⑥ respectively. Of course, the sum of β_1, β_2 and β_3 is 1. In comparison with the fuel cell (Fig. 2.1), if the effective area S is the same, the current density of the fuel cell (Fig. 2.4) is larger due to reduced effective operating area under the same operating condition. Increasing the relative humidity of air flowing into the zone ③ shall improve the performance of the fuel cell but it requires a larger area of the zone ② and ⑤. On the other hand, it is obvious that the performance of the fuel cell may be decreased at high current density for the fuel cell in Fig. 2.4 if referred to the polarization curve. This trade-off relation drives us to optimize the structure design and the system application.

Assumptions are made here for the MEA model. Firstly, all structure layers of the fuel cell are homogeneous and evenly distributed. Secondly, the local temperature and pressure are the same everywhere in the cell. Thirdly, the system is in steady state and the dimension of the model is in such direction along which the proton is conducted through the MEA. Fourthly, the distribution of air components in flow channels that are adjacent to the zone from ① to ⑥. Fifthly, the thickness of the CCL and ACL is neglected in comparison with that of the PEM, which means that thickness of the PEM is equal to that of the MEA. Sixthly, only flow of water vapor among the mass transport from zone ③ to zone ④ is considered by assuming the same diffusion characteristic of water vapor and dispersed water drop, and all gases are ideal gas.

$$\begin{cases} x_{W,Ca,In} = \dfrac{RH_{Ca}P_{sat}}{P_{Ca}}, x_{Nitro,Ca,In} = 0.79\left(1 - x_{W,Ca,In}\right) \\ x_{Oxy,Ca,In} = 0.21\left(1 - x_{W,Ca,In}\right) \end{cases} \tag{2.17}$$

In the cathode inlet, the mole fraction $x_{W,Ca,In}$ of water vapor is as Eq. 2.17, whereas $x_{Nitro,Ca,In}$ and $x_{Oxy,Ca,In}$ are the mole fraction of nitrogen and oxygen respectively, RH_{Ca} the relative humidity at the inlet, P_{sat} the saturated pressure of water vapor, and P_{Ca} the pressure in the cathode chamber.

$$\begin{cases} N_{Oxy,Ca,In} = \lambda_{Oxy}\dfrac{I}{4F}, N_{W,Ca,In} = N_{Oxy,Ca,In}\dfrac{x_{W,Ca,In}}{x_{Oxy,Ca,In}} \\ N_{Nitro,Ca,In} = N_{Oxy,Ca,In}\dfrac{x_{Nitro,Ca,In}}{x_{Oxy,Ca,In}} \end{cases} \tag{2.18}$$

$$\begin{cases} N_{Oxy,Ca,7} = N_{Oxy,Ca,In}, N_{W,Ca,7} = N_{W,Ca,In} + N_1 \\ N_{Nitro,Ca,7} = N_{Nitro,Ca,In} \end{cases} \tag{2.19}$$

$$\begin{cases} N_{Oxy,Ca,8} = N_{Oxy,Ca,In} - \dfrac{I}{4F}, N_{Nitro,Ca,8} = N_{Nitro,Ca,In} \\ N_{W,Ca,8} = N_{W,Ca,In} + N_1 + \dfrac{I}{2F} - N_2 \end{cases} \tag{2.20}$$

$$\begin{cases} N_{Oxy,Ca,Out} = N_{Oxy,Ca,In} - \dfrac{I}{4F}, N_{Nitro,Ca,Out} = N_{Nitro,Ca,In} \\ N_{W,Ca,Out} = N_{W,Ca,8} - N_3 \end{cases} \tag{2.21}$$

The nominal current density of the fuel cell is I/S but the actual value is $I/S/\beta_1$. The excess stoichiometry ratio of air and that of hydrogen are SR_1 and SR_2 respectively. The net water flux from zone ① to ② is N_1, the water flux from zone ③ to ④ $N_2 = \alpha i\beta$ and the water flux from zone ⑤ to ⑥ N_3. Based on mass conservation, the flux of gas components at the cathode inlet, the interface ⑦ and⑧, and the cathode outlet is presented as Eqs. 2.18–2.21, whereas the subscript *Oxy*, *Nitro* and *W* in variable $N_{a,b,c}$ and $x_{a,b,c}$ mean oxygen, nitrogen and water respectively, the subscript *Ca* the cathode, the subscript *In*, 8 and 7 the boundary and interface number.

$$x_{Hy,An,In} = 1 - x_{W,An,In}, x_{W,An,In} = \frac{RH_{An}P_{sat}}{P_{An}} \tag{2.22}$$

$$N_{Hy,An,In} = SR_2\frac{I}{2F}, N_{W,An,In} = N_{Hy,An,In}\frac{x_{W,An,In}}{x_{Hy,An,In}} \tag{2.23}$$

$$N_{Hy,An,9} = N_{Hy,An,In}, N_{W,An,9} = N_{W,An,In} + N_3 \tag{2.24}$$

$$N_{Hy,An,10} = N_{Hy,An,In} - \frac{I}{2F}, N_{W,An,10} = N_{W,An,In} + N_2 + N_3 \tag{2.25}$$

$$N_{Hy,An,Out} = N_{Hy,An,In} - \frac{I}{2F}, N_{W,An,Out} = N_{W,An,10} - N_1 \tag{2.26}$$

In the same way, the mole fraction of gas component at the anode inlet is as Eq. 2.22. The flux of gas components at the anode inlet, the interface ⑨ and⑩, and the anode outlet is as Eqs. 2.23–2.26, whereas the subscript *Hy* and *W* in variable $N_{a,b,c}$ and $x_{a.b,c}$ mean hydrogen and water respectively, the subscript *An* the anode, the subscript *In*, 9 and 10 the boundary and interface number. Besides, RH_{An} and P_{An} are relative humidity at the anode inlet and pressure of the anode chamber.

$$\frac{dx_{j,Ca,3}}{dy} = RT\sum_{j,k}\frac{x_{j,Ca,3}N_{k,Ca,3} - x_{k,Ca,3}N_{j,Ca,3}}{P_{Ca}D_{jk,eff}} \tag{2.27}$$

$$\begin{cases} \frac{dx_{m,An,4}}{dy} = RT\sum_{m,n}\frac{x_{m,An,4}N_{n,An,4} - x_{n,An,4}N_{m,An,4}}{P_{An}D_{mn,eff}} \\ x_{m,An,4} + x_{n,An,4} = 1 \end{cases} \tag{2.28}$$

$$P \times D_{AB} = a\left(\frac{T}{\sqrt{T_{crA}T_{crB}}}\right)^b (p_{crA}p_{crB})^{1/3}(T_{crA}T_{crB})^{5/12} \times \left(\frac{1}{M_A} + \frac{1}{M_B}\right)^{1/2} \varepsilon^{3/2} \tag{2.29}$$

The GDL at zone ③ and ④ are porous and the Stefan-Maxwell equation is applied to describe the multi-component diffusion, as Eqs. 2.27–2.28. The subscript *j* and *k* are any two of the oxygen, nitrogen and water vapor, the subscript *m* and *n* are either of the water vapor and hydrogen and $D_{jk,eff}$ and $D_{mn,eff}$ are the effective diffusion coefficient between two gas components. The gas inter-diffusion coefficient is function of the critical temperature (T_{crA} and T_{crB}), the critical pressure (p_{crA} and p_{crB}) and mole weight of gas components (M_A and M_B) and they are calculated as Eq. 2.29 after taking porosity of the GDL into consideration. In Eq. 2.29, *a* is 0.0002745 and *b* 1.832 for hydrogen, oxygen and nitrogen but *a* is 0.000364 and *b* 2.334 for water vapor.

$$\begin{cases} x_{Oxy,Ca,3} = -\frac{I}{4F}, x_{W,Ca,3} = \frac{I}{2F} - N_2 \\ x_{Hy,An,4} = \frac{I}{2F}, x_{W,An,4} = -N_2 \end{cases} \tag{2.30}$$

$$P_{Oxy,Ca,3M} = \frac{P_{Ca}x_{Oxy,Ca,3M}}{1 - x_{Liq}}, x_{Liq} = \frac{P_{Ca}x_{W,Ca,3M} - P_{sat}}{P_{Ca}} \tag{2.31}$$

The flux of gas component along the y direction in zone ③ and ④ is as Eq. 2.30 and the flux of nitrogen is zero because it does not participate in the electrochemical reaction. At the interface of zone ③ and the MEA, the activation voltage is function of the local oxygen concentration so the water content needs to be considered but in a simplified way, as Eq. 2.31. The concentration of gas components in cathode flow channel is the average of that at the interface ⑦ and ⑧ respectively. Similarly, the concentration of gas components in anode flow channel is the average of that at the interface ⑨ and⑩. As described in Sect. 2.2, the activation overvoltage is neglected at the ACL so the effect of water vapor on the hydrogen concentration at the interface of zone ④ and the MEA won't be considered.

In the membrane, the proton is conducted in combination with water molecule and the average number (the electro-osmotic drag coefficient n_{drag}) of water molecule combined with one proton is related to the local water content parameter as λ which is defined as the average number of water molecule around one perfluorosulfonic acid group. According to research by Springer [3], the electro-osmotic drag coefficient n_{drag} is equal to $2.5\lambda/22$ at 80 °C and it is assumed to be the same for temperature from 60 to 80 °C. When the local water content through the membrane is different, water diffuses towards sites where the local water concentration is lower. By neglecting the hydraulic driven diffusion of water, the local water content is governed by Eq. 2.32, whereas $D_{W,\,Mem,\,eff}$ is the effective diffusion coefficient of water in membrane as Eqs. 2.33–2.34, ρ_{dry} and $M_{mem}$the density and molecular weight of dried membrane respectively.

$$-\alpha\frac{I}{F} = n_{drag} - D_{W,Mem,eff}\frac{\rho_{dry}}{M_{Mem}}\frac{d\lambda}{dy} \tag{2.32}$$

$$D_{30,\lambda} = \begin{cases} 2.563 - 0.33\lambda + 0.0264\lambda^2 - 0.000671\lambda^3, for \lambda > 4 \\ 8 - 1.67\lambda, for 3 \le \lambda < 4 \\ 2\lambda - 3, for 2 \le \lambda < 3 \\ 1, for \lambda < 2 \end{cases} \tag{2.33}$$

$$D_{W,Mem,eff} = D_{30,\lambda}\exp\left[2416\left(\frac{1}{303.15} - \frac{1}{T}\right)\right] \tag{2.34}$$

When the local water content parameter λ is larger than 1, the local proton conductivity σ_λ in membrane is function of this parameter λ as Eq. 2.25, and the total resistance of proton conduction in membrane is as Eq. 2.36, whereas t_{Mem} is the membrane thickness. The parameter λ is calculated by Eq. 2.37 at the interface of the CGDL and MEA and that of the AGDL and MEA. It is obvious that the maximum of the parameter λ is 16.8.

$$\sigma_{\lambda} = (0.005139\lambda - 0.00326)\exp\left[1268\left(\frac{1}{303.15} - \frac{1}{T}\right)\right] \quad (2.35)$$

$$R_{Mem} = \int_{0}^{t_{Mem}} \frac{1}{\sigma_{\lambda}} \frac{dy}{S\beta_1} \quad (2.36)$$

$$\lambda_{Mem,GDL} = \begin{cases} 14 + 1.4\left(\frac{x_W P}{P_{sat}} - 1\right), for 1 \le \frac{x_W P}{P_{sat}} \le 3 \\ 0.043 + 17.81\frac{x_W P}{P_{sat}} - 39.85\left(\frac{x_W P}{P_{sat}}\right)^2 + 36\times \\ \left(\frac{x_W P}{P_{sat}}\right)^3, for 0 < \frac{x_W P}{P_{sat}} < 1 \end{cases} \quad (2.37)$$

No electrochemical reaction happens in part of the membrane sandwiched between zone ① and ② and the water diffusion inside is driven by the local water concentration. No electrochemical reaction also means that no oxygen or nitrogen or hydrogen needs to be transported in steady state in these zones. The local water content in this part of membrane is still governed by Eq. 2.32 with current density equal to zero. The mass transport equation of multi-component in the zone ③ and ④ is applicable to these zones with the flux of nitrogen, oxygen and hydrogen equal to zero. Besides, no voltage loss happens inside. This description also goes for the zone ⑤ and ⑥ and the part of membrane in between.

In addition, the concentration of gas components in the flow channel between the cathode inlet and the interface ⑦ is the average of that at these two interfaces. Similarly, the concentration of gas components in the flow channel between the cathode outlet and the interface ⑧ is the average of that at these two interfaces. The concentration of gas components in the flow channel between the anode inlet and the interface ⑨ is the average of that at these two interfaces. The concentration of gas components in the flow channel between the anode outlet and the interface ⑩ is the average of that at these two interfaces. In particular, the material at zone ① and ⑤ is highly hydrophilic and it is assumed that the water concentration at the interface between these two zones and the MEA is the same as that at the flow channel adjacent to these two zones respectively. On the other hand, the material at zone ② and ⑥ promotes water evaporation greatly so the water concentration at the interface between these two zones and the MEA is considered to be the same as that the flow channel adjacent to these two zones respectively.

Part of the operating condition and parameters of the fuel cell is enlisted (Table 2.2) and the area ratio like β_1, β_2 and β_3 is preset. Based on the water diffusion governing equation in membrane, the maximum flux of water transport is 0.224 mol/(m^{2}s) at 70 °C and that is 0.181 mol/(m^{2}s) at 60 °C. After building the model in MATLAB, the

Table 2.2 Operating condition and parameters of the fuel cell

Parameter	Value	Parameter	Value
Thickness of CGDL	250 μm	Porosity of GDL	0.288
Thickness of CCL	15 μm	RH of supplied Air	40%
Thickness of PEM	15 μm	RH of supplied H_2	40%
Thickness of AGDL	250 μm	Excess SR of H_2	2.0
Thickness of ACL	15 μm	Excess SR of air	2.0
Effective surface area	270 cm^2	Operating pressure	1.5 bar
Area Per. of ① & ②	10%	OCV	1.1 V
Area Per. of ⑤ & ⑥	10%	Operating temperature	70 °C

OCV means the open circuit voltage. SR means the stoichiometric ratio. RH means the relative humidity. Per. means the percentage

procedure for finding solution to this model is shown (Fig. 2.5) and this parameter is input as the initial condition. The 1st judging condition is whether the solution of the water content exists at the interface of the cathode gas diffusion layer and the MEA as well as the net water flux N_2. The 2nd judging condition is whether the solution of the net water flux N_3 exists.

The performance of three kinds of fuel cell is compared under the same operating condition and with the same effective area (Fig. 2.6). The first fuel cell includes three parts (Fig. 2.4). The second fuel cell includes only the second part with area ratio $1-\beta_3$ and third part with area ratio β_3. The third fuel cell includes only the second part with area ratio 1.

Based on simulation result, we can find that when the current density is lower than 0.5 A/cm^2, performance of the first fuel cell is better than that of the other two whose performance is approximately equal to each other. Furthermore, when the current ranges from 0.5 A/cm^2 to 0.65 A/cm^2, performance of the first fuel cell is equal to that of the second one but both are better than that of the third one. However, when the current density is larger than 0.65 A/cm^2, performance of the second fuel cell becomes lower than that of the third one. The higher the current density, the larger difference between the third and the second fuel cell. Interestingly, performance of the first fuel cell moves away from that of the first one and moves towards that of the third one with the increased current density. The conclusion is that the first part of the fuel cell will improve the performance of the fuel cell with no humidifier at the current density. At higher current density, the third part of the fuel cell will decrease the overall performance.

The sources of voltage loss are analyzed (Fig. 2.7) and the effective operating area of the third fuel cell is larger than that of the second one and is much larger

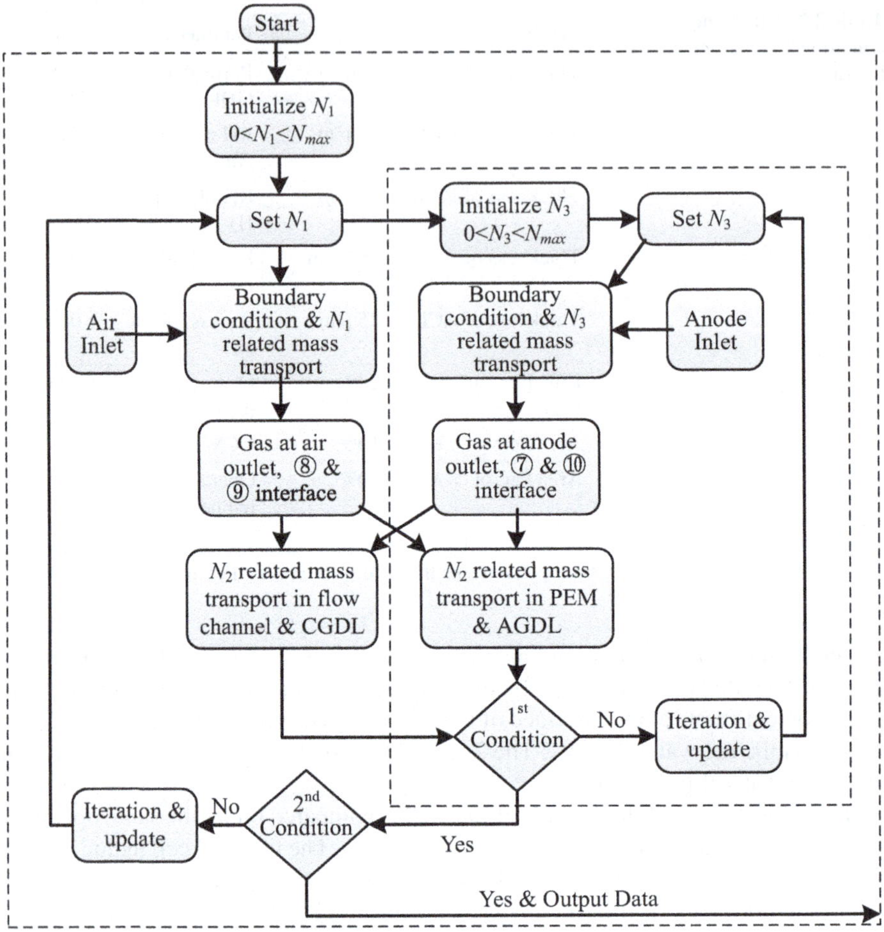

Fig. 2.5 Solution to the fuel cell model without external humidifier

Fig. 2.6 Performance of three kinds of fuel cells

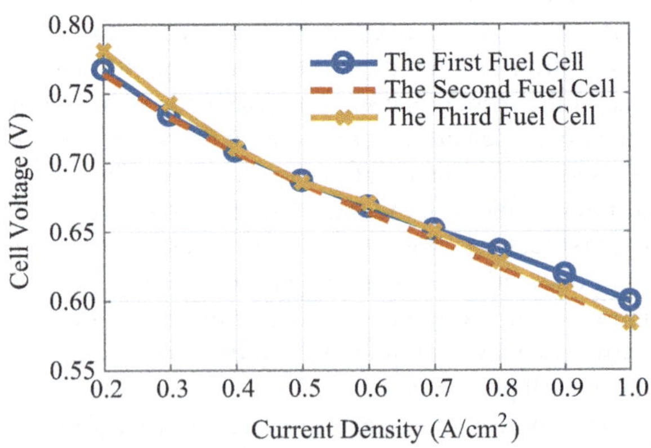

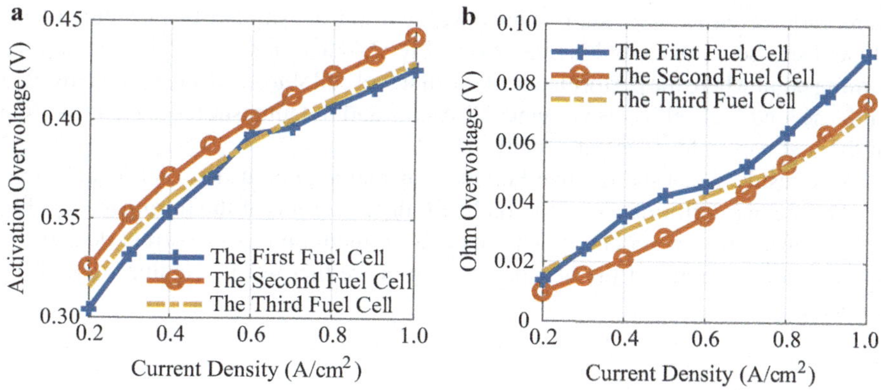

Fig. 2.7 Sources of voltage loss for the first kind of the fuel cell: **a** Ohm overvoltage, **b** activation overvoltage

than that of the first one. When the current density is lower than 0.8 A/cm^2, the ohm overvoltage of the first fuel cell is larger than that of the third one and is much larger than that of the second one. When the current density is higher than 0.8 A/cm2, the ohm overvoltage of the first fuel cell is larger than that of the second one and is much larger than that of the first one. Over the full range of current density, the activation overvoltage of the first fuel cell is lower than that of the third one and much lower than that of the second one.

Here, the water flux through the first part or the third part divided by the water flux through the second part is defined as the contribution of water recirculation in either part. The analysis result is presented (Fig. 2.8). For the first fuel cell, the contribution of water recirculation through the first part increases gradually with current density but when the current density is higher than 0.7 A/cm^2, it decreases gradually with current density. On the contrary, the contribution of water recirculation through the

Fig. 2.8 Water flux contribution and current density

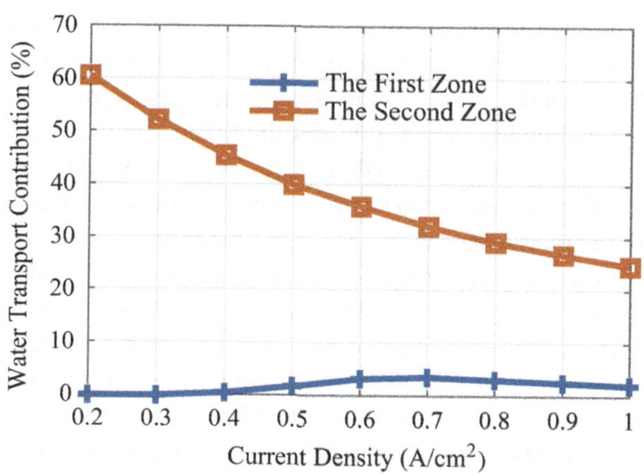

third part decreases more significantly with current density but the absolute of the slope decreases. If we combine the result (Fig. 2.6) and that (Fig. 2.7) together, the turning point where the performance of the first fuel cell starts to decrease towards that of the second fuel cell is also around 0.7 A/cm^2 and it demonstrates the contribution of the first part in low current density.

As a result, when the relative humidity of fuel supply at the stack inlet is low, to enhance the performance of the fuel cell, the area ratio of the third part ought to be decreased. If the fuel cell mostly operates at medium or low current density, an appropriate area ratio of the first part will improve the overall running state of the fuel cell.

2.4 Modeling of Water Content Inside the MEA

The micro structure of the catalyst layer in MEA is very complex and the micro-pore inside forms the path where the water vapor and oxygen are partially transported. However, the micro-pore is not in the cylinder shape and it is tortuous and twisted. To make it easier to understand its principle, the scheme is presented in view of one micro-pore on both sides of the membrane along the direction in which the proton is conducted (Fig. 2.9). The effective porosity is defined as the void area divided by the total area of the catalyst layer at any section which is perpendicular to the proton conduction direction. It is obvious that when liquid water accumulates in the micro-pore, the effective porosity is decreased while the local proton conductivity is enhanced due to more water molecule around one perfluorosulfonic acid group.

Research shows that when the catalyst layer is full of liquid water, the effectively-utilized thickness of the catalyst layer is decreased significantly with the increased current density. Just one case is that the ratio of effectively-utilized thickness is less than 5% at current density higher than 0.5A/cm^2 and it indicates that the thickness of the catalyst layer is increased largely where only proton is conducted but no electrochemical reaction like oxygen reduction happens.

At the interface of the membrane and CCL, the local water content parameter reaches its maximum through the membrane because the net water flux is from the cathode to the anode, when the hydrogen at the anode is dry and the air at the cathode is partially humidified. The activation overvoltage at the ACL can be neglected compared to that at the CCL, so the ACL is treated as an extension of the membrane and the governing equation of water transport inside is the same as that in the membrane as in Sect. 2.2. Our research will be focused on the CCL.

The cathode catalyst layer is composed of the carbon agglomerate, the platinum agglomerate, the PTFE, the polymer electrolyte, water and other additives (Fig. 2.10). The carbon agglomerate provides the path for electronic conduction and acts as substrate to bear the platinum agglomerate. The platinum agglomerate is very small to have large specific surface area. The PTFE connects all carbon agglomerate together to form the overall structure with mechanical property. The polymer electrolyte

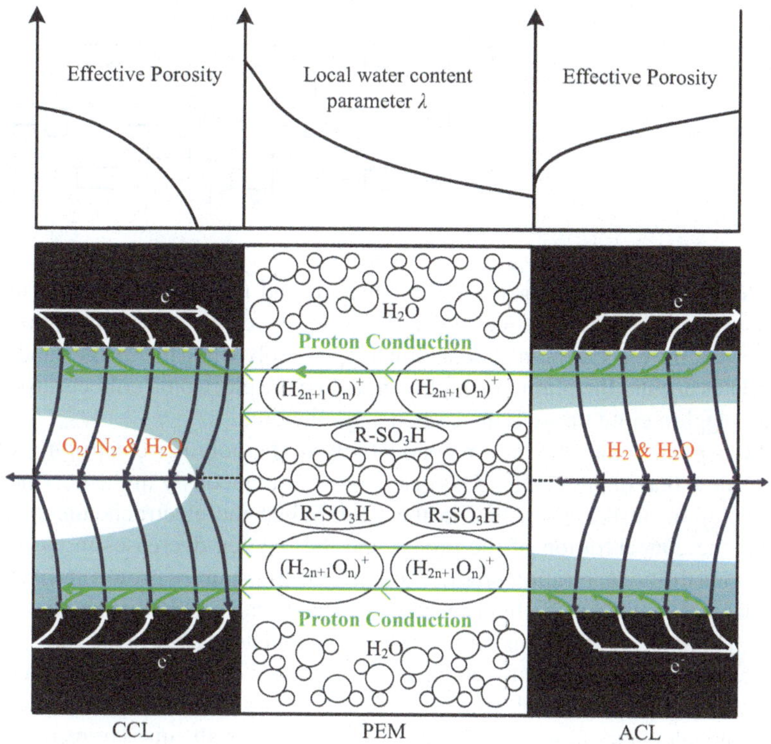

Fig. 2.9 Principle of the MEA

Fig. 2.10 Mass transport in the cathode catalyst layer

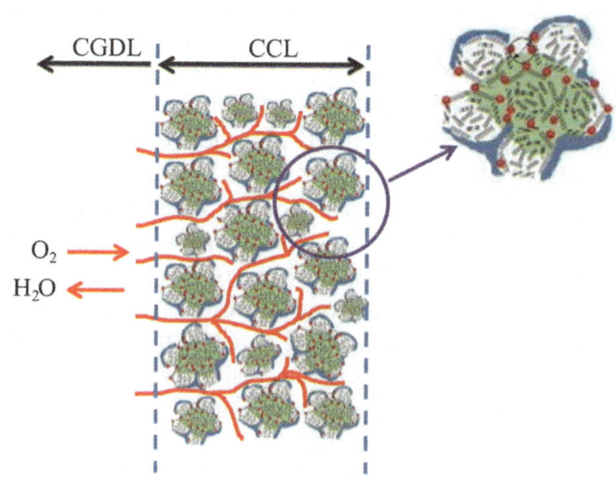

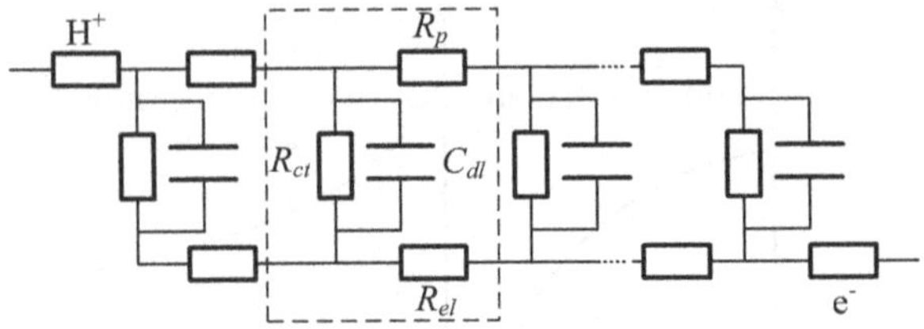

Fig. 2.11 Transmission line model

and water forms the path for proton conduction and the electrochemical reaction happens where the polymer electrolyte and water cover the platinum agglomerate. That is to say, the electrochemical reaction happens only when the oxygen and proton accumulates around the platinum agglomerate and the electronic collects around the platinum agglomerate through the carbon agglomerate.

To be noticed is that the proton conductivity in the polymer electrolyte solution is higher than that in pure water. Oxygen exists simultaneously in the polymer electrolyte solution, in the water and in the void. When the electrochemical reaction happens, the concentration of the proton and the oxygen decreases in the site and then it forms the concentration gradient to drive the transport of proton and oxygen to the site from the outside. The byproduct water accumulates in the site and the concentration gradient drives the transport of water to the outside either in the form of water vapor or liquid water or both together. The inner space of the carbon agglomerate is filled with solution or just solvent.

The equivalent circuit model (Fig. 2.3) cannot describe the mass transport, charge conduction and the electrochemical reaction completely, so Eikerling proposed the transmission line model (Fig. 2.11), in the approximate finite element analysis mode [4]. To be specific, R_p is the resistance of proton conduction, R_{el} the resistance of electronic conduction, R_{ct} the Faradic resistance at the electrochemical reaction site, and C_{dl} the double layer capacitor at the electrochemical reaction site. When the number of the unit in the gray-background zone tends towards infinity, this equivalent circuit model can quantify the physical-chemistry process in the catalyst layer more exactly.

Considering the distribution of water content in the MEA and its influence on performance of the MEA, the lumped model for the catalyst layer including the Butler-Volmer equation and the voltage loss resulting from the proton conduction in this layer is not sufficient. Therefore, based on principle of the catalyst layer, the modeling of the MEA is elucidated as follows and please refer to the Fig. 2.12 for details. The micro-pore in the catalyst layer is assumed to be of the same length and the same diameter for simplification.

The direction of the electronic conduction is defined as the positive direction of coordinate x. The coordinate x starts from the interface of the CCL and CGDL. The coordinate x ends at the interface of the CCL and membrane and the thickness of the CCL is l.

The proton moves from the interface where x is l towards the interface where x is 0 and the electronic moves from the interface where x is 0 towards the interface where

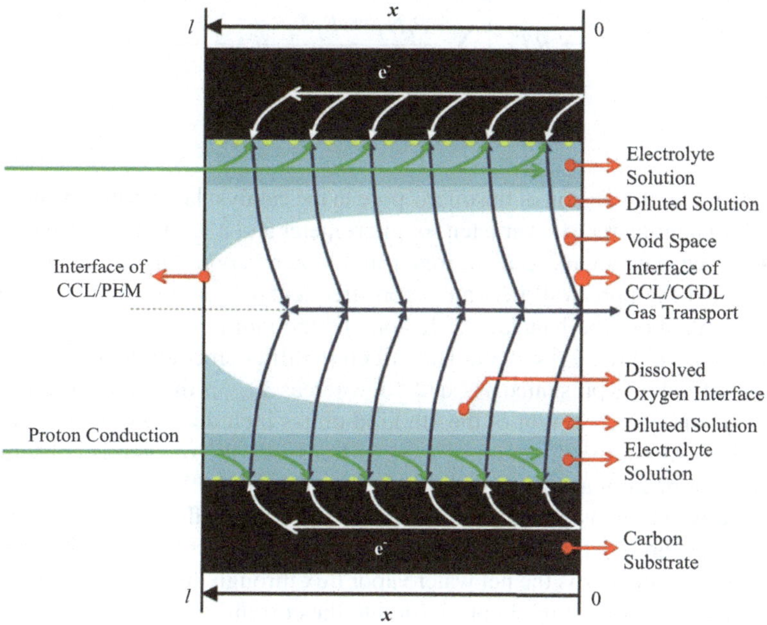

Fig. 2.12 Principle of the cathode catalyst layer

x is l. The proton is conducted in the polymer electrolyte solution but the electronic is conducted in the carbon substrate. In the void space at the micro-pore center, oxygen is transported from the interface where x is 0 towards the interface where x is l. In the void space at the micro-pore center, water vapor diffuses from the interface where x is l towards the interface where x is 0. Nitrogen exists in all space of the micro-pore but it doesn't take part in the reaction. To maintain the electrochemical reaction, the oxygen must be dissolved into the water, then the dissolved oxygen is transported through the water and the polymer electrolyte solution and finally it reaches the reaction site.

Assumptions are made here for simplification as follows. Firstly, the substance in the catalyst layer is homogeneous for each phase. Secondly, the temperature and the pressure is the same for space in the whole catalyst layer. Thirdly, the model is pseudo-two dimensional because the length of the micro-pore is far larger than its diameter. Fourthly, the MEA is in stable state. Fifthly, along the length of the micro-pore, the diameter of the micro-pore, the thickness of the polymer electrolyte solution and water over the internal wall of the micro-pore are constant. Sixthly, the proton conductivity is the same for the polymer electrolyte solution and the electronic conductivity the same for the carbon substrate. At last, the membrane is still governed by model proposed by Springer.

$$\frac{dP_i}{dx} = RT_{fc}\left[\sum_j \frac{P_iN_j - P_jN_i}{P_{CCL}D_{ij}} - \frac{N_i}{D_{i,kdeff}}\right] \tag{2.38}$$

$$P_W + P_{Oxy} + P_{Nitro} = P_{CCL} \tag{2.39}$$

The diameter dimension of the micro-pore in the catalyst layer ranges from several hundreds of nanometer to several tens of micrometer and it's smaller than the average molecule free path of oxygen, nitrogen and water vapor. The collision between molecule and the pore wall becomes more frequent so gas transport in the micro-pore must take both the Knudsen diffusion and the multi-component diffusion into consideration, as Eqs. 2.38–2.39. The effective diffusion coefficient $D_{i,kdeff}$ of the Knudsen diffusion is presented as Eq. 2.40, whereas d_{pore} is the micro-pore diameter and M_k the molecular weight of the kth kind of gas including oxygen, nitrogen and water vapor. Along the direction of oxygen transport, the oxygen is consumed by the electrochemical reaction so its flux N_{Oxy} is decreased all the way ever since the oxygen starts to be consumed, as Eq. 2.41. The water vapor flux N_w is function of the net water flux through the membrane and the oxygen flux, as Eq. 2.42. The net water vapor flux coefficient α is the net water vapor flux through the membrane divided by the generated water per time in proportion to the current.

$$D_{i,kdeff} = \frac{d_{pore}}{3}\sqrt{\frac{8RT_{fc}}{\pi M_k}} \tag{2.40}$$

$$\frac{dN_{Oxy}}{dx} = -N_{csp} \tag{2.41}$$

$$N_W = -2N_{Oxy} - \alpha\frac{I}{2F} \tag{2.42}$$

The concentration c_{Oxy} of dissolved oxygen at the interface of the void space and polymer electrolyte solution is quantified as Henry Law and it's proportional to the oxygen partial pressure, as Eq. 2.43, whereas $c_{Oxy,0}$ is the standard concentration of dissolved oxygen at standard atmosphere pressure.

$$c_{Oxy} = P_{Oxy}c_{Oxy,0} \tag{2.43}$$

$$N_{csp} = \alpha_v D_{Oxy,Liq}\frac{c_{Oxy} - c_{Oxy,Carb}(r_{pore}, x)}{\delta_{Liq}} \tag{2.44}$$

$$N_{csp} = \varepsilon_c K_{react}C_{Oxy,Carb}(r_{pore}, x) \tag{2.45}$$

The thickness of the polymer electrolyte solution over the micro-pore wall is very small and the transport of the dissolved oxygen in the polymer electrolyte solution is governed by the Fix Diffusion Law. The local consumption rate N_{csp} of oxygen is presented as Eq. 2.44, whereas α_v is the specific surface area per unit volume, $D_{Oxy,Liq}$ the effective diffusion coefficient in the solution, δ_{Lid} the thickness of solution film over the micro-pore wall, $C_{Oxy,Carb}$ the local oxygen concentration on the surface of the platinum agglomerate [5]. The consumption rate of oxygen is related to the electrochemical reaction rate K_{react} and the local oxygen concentration $C_{Oxy,Carb}$, as Eq. 2.45. The effective factor ε_c is function of the corrected first class Bessel function, as Eqs. 2.46, 2.47 and 2.48, whereas r_{ag} is the average radius of the carbon substrate agglomerate and $E_n E_{agg}$ is the coefficient for internal space in the agglomerate [5].

$$\varepsilon_C = \frac{I_1(2M_C)}{M_C I_0(2M_C)} \tag{2.46}$$

$$M_C = \frac{r_{agg}}{2}\left(\frac{K_e}{E_n E_{agg} D_{Oxy,Liq}}\right)^{1/2} \tag{2.47}$$

$$\varepsilon_C = \frac{3M_C \coth(3M_C) - 1}{3M_C^2} \tag{2.48}$$

Based on Eqs. (2.44) and (2.45), the unknown variable $C_{Oxy,Carb}(r_{pore},x)$ can be neglected and we get a new Eq. (2.49). According to the Butler-Volmer equation, the local electrochemical reaction rate N_{csp} is function of the electrode over-potential as Eq. (2.50). Then the electrochemical reaction rate K_{react} is calculated as Eq. (2.51) [5].

$$N_{csp} = \frac{c_{Oxy}}{\frac{\delta}{\alpha_v D_{Oxy,Liq}} + \frac{1}{\varepsilon_c K_e}} \tag{2.49}$$

$$\frac{dI_p(x)}{dx} = \alpha_v i_{exe} \frac{c_{Oxy,Carb}}{c_{Oxy,ref}} \times \left[\exp\left(\frac{\alpha_{An} F \eta_{op}}{RT_{fc}}\right) - \exp\left(\frac{-\alpha_{Ca} F \eta_{op}}{RT_{fc}}\right)\right] \tag{2.50}$$

$$K_e = \frac{\alpha_v i_{exe}}{4F c_{Oxy,ref}} \tag{2.51}$$

The electrode over-potential is the potential of the solid medium minus the potential of the polymer electrolyte solution and the electrode over-potential for the electrochemical reaction at the CCL is negative, as Eq. 2.52. The current in the catalyst layer and along the x direction is the same as the output current of the fuel cell and it's sum of the electronic-conduction current I_e and the proton-conduction current I_p, as Eq. 2.53. In the catalyst layer, the electronic is conducted in the solid medium while the proton is conducted in the polymer electrolyte solution. The any-type charge conduction will result in variation of voltage along the x direction, as Eq. 2.54.

$$\eta_{op} = \eta_s - \eta_{eles} \tag{2.52}$$

$$I = I_p(x) + I_e(x) \quad (2.53)$$

$$\begin{cases} \frac{d\eta_{eles}}{dx} = \frac{I_p(x)}{\sigma_{p,CCL}} \\ \frac{d\eta_s}{dx} = \frac{I_e(x)}{\sigma_e} \end{cases} \quad (2.54)$$

Along the x direction, the relation between oxygen flux N_{Oxy} and the electronic-conduction current I_e is presented as Eq. (2.55). Combining Eqs. 2.52–2.55 together, the local electrode over-potential is derived as Eq. 2.56 along the x direction. The boundary condition at two interfaces and some constraints are presented as Eqs. 2.57–2.59.

$$I_e(x) = 4FN_{Oxy} \quad (2.55)$$

$$\frac{d\eta_{op}}{dx} = \frac{I}{\sigma_{p,CCL}} - 4FN_{Oxy}\left(\frac{1}{\sigma_{p,CCL}} + \frac{1}{\sigma_e}\right) \quad (2.56)$$

$$I_e|_{x=0} = I,\ I_p\big|_{x=0} = 0,\ N_{Oxy}\big|_{x=0} = \frac{I}{4F},\ N_W|_{x=0} = -\frac{(2\alpha + 1)I}{2F} \quad (2.57)$$

$$I_e|_{x=l} = 0,\ I_p\big|_{x=l} = I,\ N_{Oxy}\big|_{x=l} = 0,\ N_W|_{x=l} = -\alpha I \quad (2.58)$$

$$I_e(x) \geq 0,\ I_p(x) \geq 0,\ N_{Oxy}(x) \geq 0,\ N_W(x) \leq -\alpha I \quad (2.59)$$

In the catalyst layer, oxygen is transported from the interface of the CGDL and CCL to the interface of the CCL and membrane. Any electrochemical reaction at x will result in the gradient of oxygen concentration and no reaction means no oxygen flux. If no reaction happens at the position x_0, no reaction happens in the zone from position x_0 to the membrane and the effectively-utilized thickness is just x_0.

The parameters for the fuel cell and the operating conditions are listed (Table 2.3) and other parameters are referred to Table 2.2. The model is set up and it's simulated

Table 2.3 Operating condition and parameters of the MEA

Parameter	Value	Parameter	Value
Avg. diam. of pore	70 nm	Reference oxy. conc	1.2 mol/m^3
Proton cond. in CL	3.052 (Ω·m)$^{-1}$	Oxy. diff. coeff. in electr	6.7*10^{-10} m^2/s
Electronic cond. in CL	88.6 (Ω·m)$^{-1}$	Porosity of CL	0.3
Diam. of carbon agglomerate	500 nm	Ratio of electr. in solid Pha	0.3
Anodic charge transf. coeff	0.5	Cathodic charge transf. coeff	1.0

Avg. means the average. Cond. means the conductivity. Diam. means the diameter. Transf. Coeff. means the transfer coefficient. Oxy. Conc. means the oxygen concentration. Electr. means the electrolyte. Pha. Means the phase of substance in terms of the current density, activation overvoltage and the water content. For example, when the current density is 0.8 A/cm^2 and the net water flux coefficient is -0.1, the activation overvoltage is increased and the effectively-utilized thickness of the catalyst layer is decreased as the turning point where the oxygen flux is zero, if the electrode over-potential at the interface of the cathode gas diffusion layer and the cathode catalyst layer is increased (Fig. 2.13). At the same time, the zone is enlarged where just the proton is conducted and water is transported but no reaction happens (Fig. 2.13). That is to say, the water content in the catalyst layer is increased. The same situation goes for difference current density and net water flux coefficient.

When the current density is 0.8 A/cm^2 and the net water flux coefficient is -0.1, the proton conduction resistance in the MEA is increased with the increase of the electrode over-potential at the interface of the CGDL and CCL (Table 2.4). To be noticed is that the local water content parameter at the interface of the membrane and the cathode catalyst layer is assumed to keep the maximal and if the net water flux coefficient is constant, the proton-conduction resistance in the membrane is constant. This result does not mean that the increase of water content in the MEA will result in increase of the proton-conduction resistance of the MEA. Our focus is partially

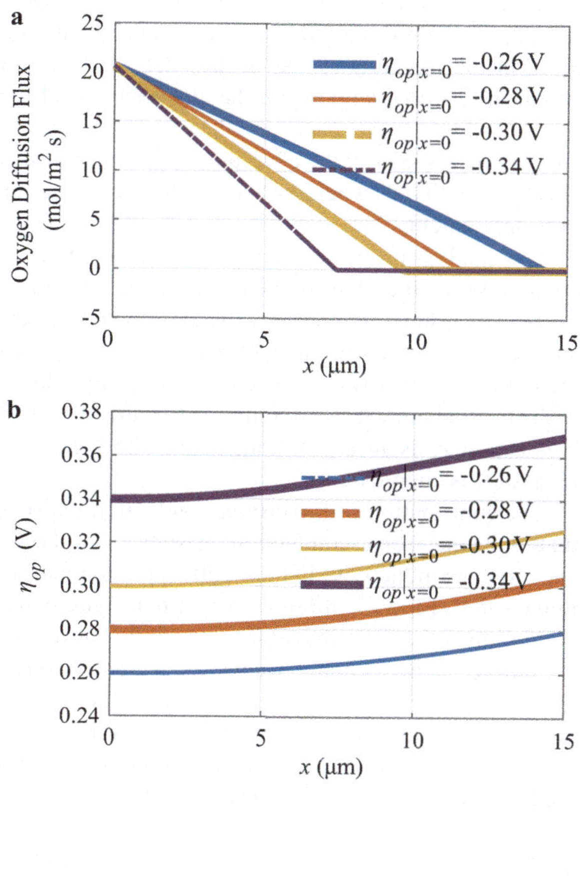

Fig. 2.13 **a** Oxygen diffusion flux with respect to distance, **b** potential difference between solid phase and solution

Table 2.4 Analysis of the proton-conduction resistance

PD at CGDL/CCL (V)	PCR in CL (mΩ·cm²)	PCR in PEM (mΩ·cm²)	Total PCR (mΩ·cm²)
−0.26	2.4	29.8	32.2
−0.28	11.1	29.8	40.9
−0.30	17.6	29.8	47.4
−0.34	24.9	29.8	54.7

PD means the potential difference between the solid phase and the solution. PCR means the proton conduction resistance

studying proton conduction resistance in the catalyst layer and it cannot be neglected with increased water content.

The simulation results tell that the activation overvoltage in the MEA and the proton-conduction resistance in the catalyst layer where no reaction happens are the two important parameters for quantification of water content in the MEA.

Based on the water content model of MEA and the transmission line model, the complete equivalent circuit model of the MEA is illustrated as Fig. 2.14. The difference between the transmission line model and our equivalent circuit model is as follows: the proton-conduction resistance including the proton conduction resistance in the membrane, that in the ACL, and that in the CCL where no electrochemical reaction happens; in the electrochemical reaction zone, the faradic resistance, the double-layer capacitor, the proton-conduction resistance and the electronic-conduction resistance; the electronic-conduction resistance only in the catalyst layer close to the CGDL where no electrochemical reaction happens as well due to lack of water content.

The catalyst layer can be divided into three parts and it depends on the water content. When the catalyst layer is dried out, the zone with electrochemical reaction moves towards the membrane, the zone with only proton-conduction resistance is suppressed and the zone with only electronic-conduction resistance is enlarged. On the contrary, when water content in the catalyst layer is being increased, the zone with electrochemical reaction moves towards the CGDL, the zone with only proton-conduction resistance is enlarged and the zone with only electronic-conduction resistance is suppressed.

The contribution of the proton-conduction resistance is significantly larger than that of the electronic-conduction resistance and it results from the significant difference in the conductivity in their corresponding medium. The activation overvoltage in the catalyst layer is minimal only when water content is in appropriate range.

In terms of the AC impedance, when the current excitation signal is imposed on the fuel cell at high enough frequency, the double-layer capacitor is short circuit and the equivalent circuit model is equal to sum of the proton-conduction resistance in membrane and the anode catalyst layer, that in the only proton-conduction zone and the electronic-conduction resistance in the only electronic-conduction zone. This is the high frequency resistance (HFR) in MEA.

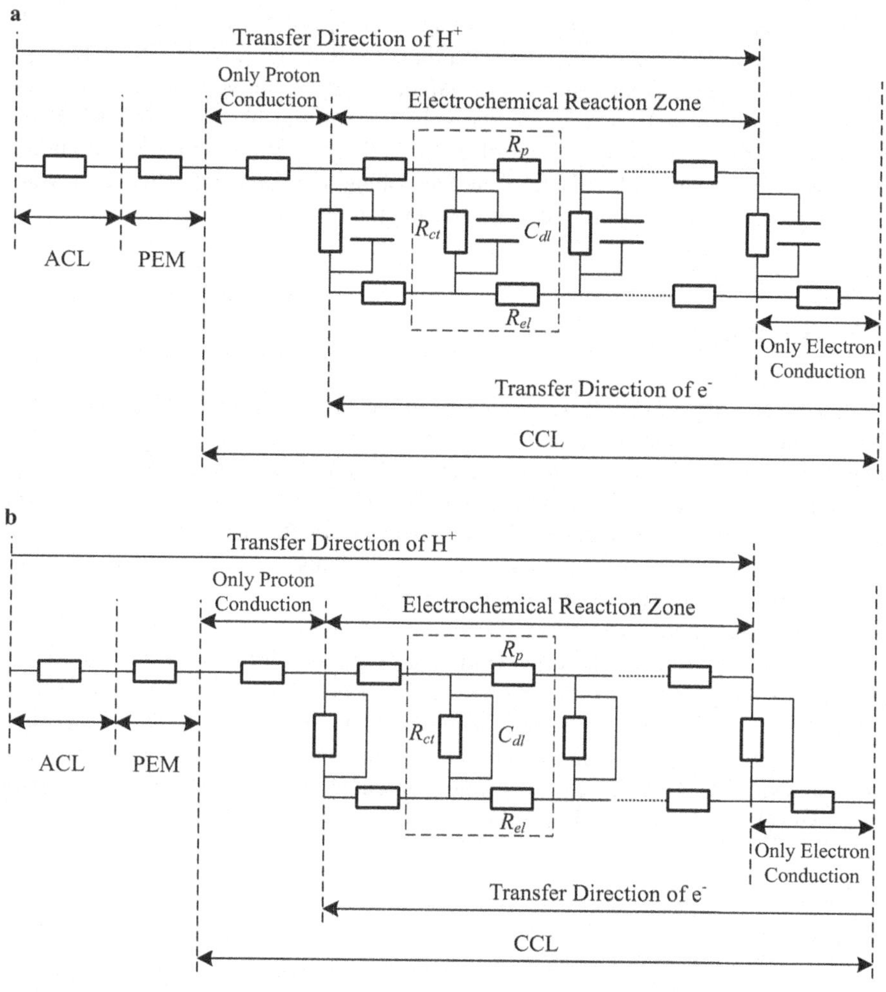

Fig. 2.14 **a** Complete equivalent circuit model of the fuel cell, **b** the high frequency resistance of the fuel cell

If the electronic conductivity in the carbon substrate varies little under relative stable operating conditions, the variation of the measured HFR is dominated by the proton conduction related to water content. Moreover, the contribution of the proton-conduction resistance to the overall HFR of MEA cannot be neglected and treating the proton conduction in the catalyst layer as just extension of that in the membrane is not reasonable.

For a fuel cell, the activation overvoltage can be obtained in two ways: firstly, the open circuit voltage of the fuel cell minus the voltage resulting from the high frequency resistance is an approximate value; secondly, the voltage resulting from the low frequency resistance minus the voltage resulting from the high frequency resistance is the more accurate one. If the high frequency resistance is measurable for a fuel cell system and the activation overvoltage is obtainable as well, the water content in the fuel cell can be estimated online based on the MEA model.

2.5 Summary

In this chapter, the model of water content in the MEA of a fuel cell is established. Firstly, the polarization curve and the equivalent circuit model is proved to be mathematically equivalent and the relation between electrical components and the voltage loss resources are sorted. The high frequency resistance reflects mainly the proton-conduction resistance and the low frequency resistance describes the overall voltage loss. The voltage loss by the high frequency resistance minus the low frequency resistance is certainly the activation overvoltage.

Secondly, a model for special structure of a fuel cell operated under no-humidifying conditions is set up to study the internal recirculation of water content and it tells the potential to improve the fuel cell performance in low and medium current density. Finally, the water content model for MEA tells that the proton-conduction resistance and the activation overvoltage together can quantify the water content inside MEA and the effect of water content on performance and characteristic of a fuel cell can be reflected in variation of the complete equivalent circuit model.

References

1. Fujitani H (2008) Fuel cell: Japan, 2008140636A. 2008–07–19
2. Origasa Y, Ando M (2010) Fuel cell: Japan, 2010009780A. 2010–01–14
3. Springer TE, Zawodzinski TA, Gottesfeld S (1991) Polymer electrolyte fuel cell model. J Electrochem Soc 138(8):2334–2342
4. Eikerling M, Kornyshev AA (1998) Modelling the performance of the cathode catalyst layer of polymer electrolyte fuel cells. J Electroanal Chem 453(1–2):89–106
5. Raymond PI, Michael BC (1980) Voltage losses in fuel cell cathodes. J Electrochem Soc 127(7):1433–1440

Chapter 3
Approach to Online AC Impedance Measurement of the Fuel Cell

3.1 Introduction

The theme of this chapter is to find the approach to online AC impedance measurement of the individual fuel cell and the stack when the fuel cell system is applied to the vehicle. In the fuel cell system, only the DC/DC converter is connected directly to the fuel cell stack and integrating the function of online AC impedance measurement into this converter is a good choice for reasons. Firstly, the DC/DC converter is an actively-controllable device to adjust the input voltage to the output voltage requested by the powertrain system or the corresponding current. Secondly, the DC/DC converter needs to measure the input voltage and current of the stack and this is also required by signal acquisition for AC impedance calculation. Thirdly, the fuel cell system needs a function without more parts but the cost is acceptable. Besides, our target is to analyze the principle and apply this approach to more situations.

Based on the principle of the electrochemical impedance spectroscopy common in the electrochemical research, three important steps are summarized as follows. Firstly, the current/voltage excitation needs to be superimposed on the stable output of the fuel cell stack and then the fuel cell and the stack show responses with alternative voltage/current respectively. Secondly, appropriate sensors for voltage and current and the signal processing circuit work together to simultaneously extract the excitation signal and the corresponding response signal. Thirdly, the microprocessor samples the processed signals and using the FFT technique to calculate the impedance in terms of amplitude and phase. This provides effective guidance for our research.

The content of this chapter is organized as follows: in Sect. 3.2, the powertrain system powered by the fuel cell stack is analyzed and two proposals for injecting the excitation to the fuel cell stack is put forward; in Sect. 3.3, the two proposals are modeled theoretically and analyzed based on the simulation result to prove the feasibility of them; in Sect. 3.4, the weak signal extraction method with high speed and high accuracy is proposed to obtain target signals; in Sect. 3.5, experiment is

P. Hong, *Water Content Estimation and Control of PEM Fuel Cell Stack and the Individual Cell in Vehicle*, Springer Theses, https://doi.org/10.1007/978-981-16-8814-0_3

performed on the lithium-ion battery pack and the fuel cell stack to verify feasibility of our proposals; in Sect. 3.6, research in this chapter is summarized.

3.2 The Specialized DC/DC Converter for the Fuel Cell Stack

The hybrid powertrain system of the fuel cell vehicle has three common topologies. The first topology is put forward for the commercial vehicle by research group of Tsinghua University (Fig. 3.1). The latter two topologies are put forward by Toyota Motor Corporation for the passenger vehicle and they have been applied to the vehicle released in 2008 and the vehicle MIRAI released in 2014 (Fig. 3.2) [1]. Report in patents and the SAE conference paper shows that the third topology (Fig. 3.2b) has been adopted by Nissan Corporation, Hyundai Corporation, Honda Corporation and Ford Corporation but the parameter configuration is different.

These topologies have both advantages and disadvantages. In the first topology, the fuel cell stack is connected to the input of the DC/DC converter and the output voltage of the battery pack is shared by the output of the DC/DC converter and the input of the inverter for traction motor. The capacity of the battery pack is large as well as the voltage to ensure the high efficiency of the motor. The fuel cell system can operate in stable state and it's helpful for the durability of the fuel cell stack.

In the second topology, the battery pack is connected to input of the bidirectional DC/DC converter and the output voltage of the fuel cell stack is shared by the output of this converter and the input of the inverter for traction motor. The voltage of the battery pack is matched using this converter. The fuel cell system must be of high voltage and fast dynamic response while the battery pack is of small capacity and low voltage. The dynamic power request from the traction motor deteriorates performance and durability of the fuel cell system and until now this topology has been the history.

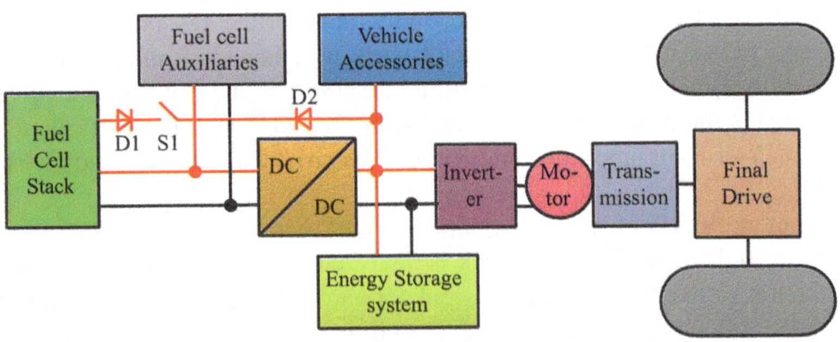

Fig. 3.1 Powertrain system of the fuel cell vehicle by Tsinghua University

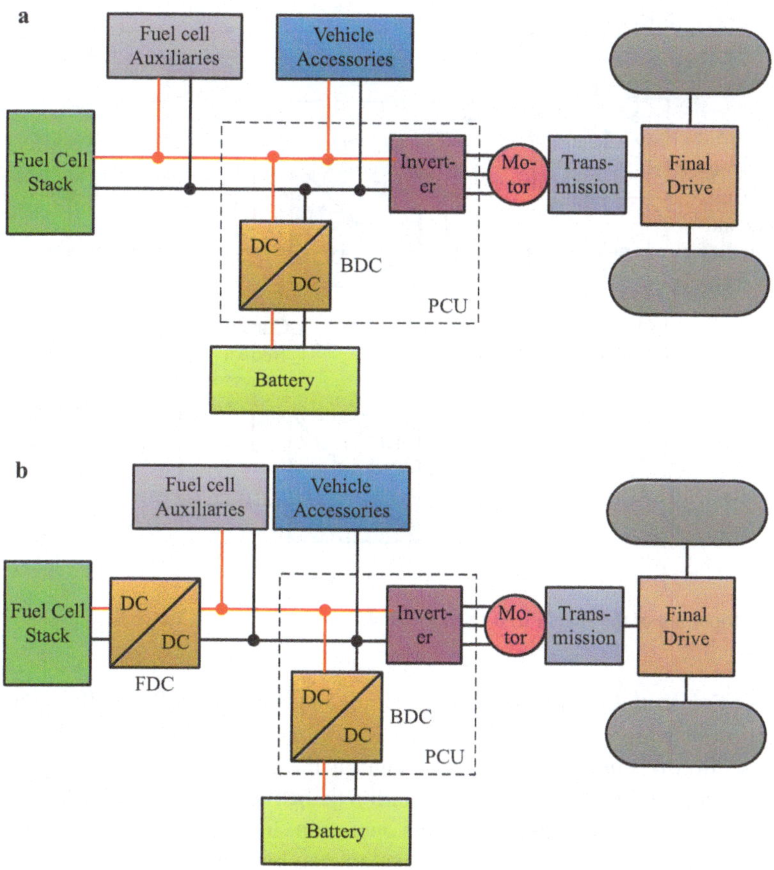

Fig. 3.2 Powertrain system of the fuel cell vehicle by Toyota Motor Corporation: **a** The first generation, **b** The second generation [1]

In the third topology, the fuel cell stack is connected to input of the first DC/DC converter and the battery pack connected to input of the second bidirectional DC/DC converter. The bus voltage is shared by output of the first DC/DC converter, output of the second bidirectional DC/DC converter and the input of the inverter for traction motor. The battery pack is of low voltage and low capacity. The power request is mostly satisfied by the fuel cell system and the transient power request can be satisfied by the battery pack, which contributes to performance and durability of the stack. In fact, the system control is complex and the risk of overshoot of the bus voltage emerges.

Either in the first or in the third topology, the fuel cell system is connected to a DC/DC converter so this converter is the research object for online AC impedance measurement of the individual fuel cell and the stack.

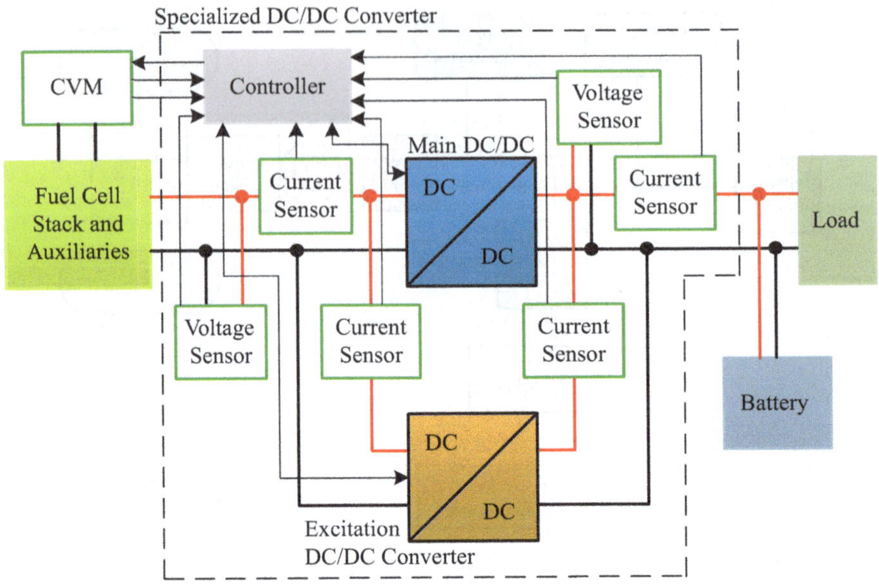

Fig. 3.3 Topology of the specialized DC/DC converter

According to the polarization curve (Fig. 2.2), the fuel cell usually operates in the section of which the slope is smaller in comparison with that of other section. In this section of curve, the output voltage of the fuel cell varies slowly with the increased current. Injecting voltage excitation to the fuel cell will lead to large current fluctuation, which is not appropriate for operation. Instead, the current excitation is our first choice and the linearity of the system can be met.

To superimpose the current excitation on the stable output current of the fuel cell, it can be realized theoretically in way of two shunt circuits in parallel and two topologies are given for consideration (Figs. 3.3 and 3.4).

The specialized DC/DC converter is composed of two DC/DC converters (Fig. 3.3). The first DC/DC converter, as DC1, is controlled to draw constant current from the fuel cell stack and the nominal power of this converter is matched with the powertrain system. The input current of the second DC/DC converter, as DC2, is controlled to generate current excitation. Both input of DC1 and DC2 is connected in parallel and both output of DC1 and DC2 connected in parallel. Both converters convert the input voltage to the output voltage with energy loss. That is to say, this topology decouples the power control and the generation of current excitation, and the system is more flexible.

On one hand, if the requirement for system integration is not strict, a new DC/DC converter can be connected to the original DC/DC converter as this topology to realize the function. On the other hand, if the system has strong desire for small volume and low weight, DC1 and DC2 shall be integrated as one DC/DC converter to simplify the design for heat dissipation, EMC and high voltage connectors and so on. Usually,

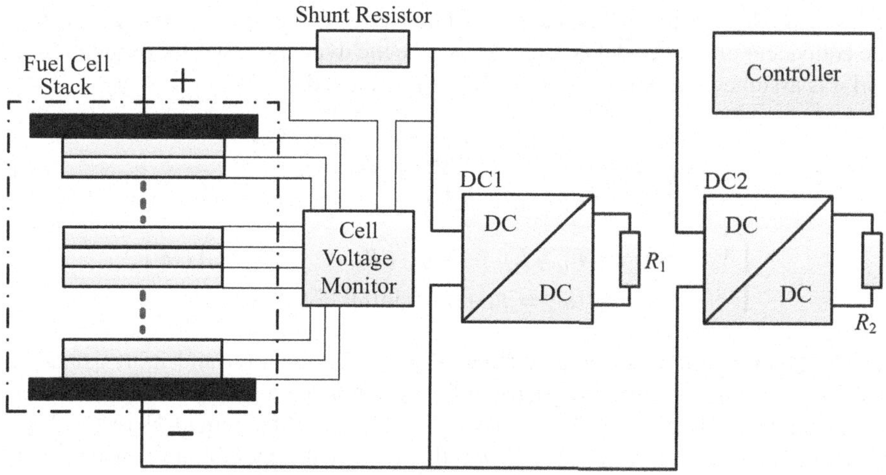

Fig. 3.4 Second topology of the DC/DC converter

the output current of the fuel cell system is as large as several hundred amperes and to improve efficiency, the interleaved multi-phase topology is utilized to lower current in each phase. The actually used phase depends on the requested power and only when the power is close to the peak power, all phases are activated. For normal operation at low power, some phases are available for other application and this is just for reference. If so, the integration is kind of feasible and more importantly, this topology is focused on the parallel connection in function but it doesn't necessarily mean two separate DC/DC converters in real.

The topology is different from the specialized DC/DC converter because output of DC1 and DC2 is separate (Fig. 3.4). The output of DC1 is to be connected to both output of the battery pack and input of inverter for traction motor, while the output of DC2 can be connected to a resistor to convert electricity to heat or it can provide power to auxiliaries in the fuel cell system. In this way, the design of DC2 is more flexible to adapt to various situations. Even though the energy loss emerges for this topology, the online AC impedance is not activated through operation of the fuel cell system and the energy loss is controllable.

3.3 Modeling and Simulation for Feasibility Analysis

In this section, the focus is on the feasibility analysis of the two topologies (Figs. 3.3 and 3.4), in terms of the electrical characteristic of the fuel cell stack with DC/DC converter. To begin with, the output voltage of the fuel cell stack is lower than the required voltage of the inverter for traction motor so the DC/DC converter operates in boost mode and the non-isolated boost converter is the research object. Besides, the equivalent circuit model is used to describe the dynamic response of the fuel cell and it is assumed that uniformity of cells is guaranteed.

$$\begin{cases} V_{fc}(\omega) = E_N - \left(R_m + \frac{R_F}{j\omega C_{dl} R_F + 1}\right) I(\omega) \\ I(\omega) = I_0 + I_{AC}\sin(\omega t + \theta_0) \end{cases} \tag{3.1}$$

$$\begin{cases} V_{stack}(\omega) = N_{stack} E_N - N_{stack}\left(R_m + \frac{R_F}{j\omega C_{dl} R_F + 1}\right) I(\omega) \\ I(\omega) = I_0 + I_{AC}\sin(\omega t + \theta_0) \end{cases} \tag{3.2}$$

The relation between output voltage $V_{fc}(\omega)$ of a fuel cell and its resistance is presented as Eq. 3.1, and the relation of output voltage V_{stack} of the stack and its resistance presented as Eq. 3.2, whereas E_{Nernst} is the theoretical Nernst voltage, N_{stack} number of cells in the stack, R_m the sum of the membrane resistance, the electronic conduction resistance and the contact resistance, R_F and C_{dl} the Faradic resistance and double-layer capacitor of the electrochemical reaction at catalyst layer respectively, I the output current, I_0 the stable output current, I_{AC}, ω and θ_0 amplitude, angular frequency and initial phase of the current excitation respectively.

$$\begin{cases} V_{stack} = N_{stack} E_N - N_{stack}(R_m + R_F) I_0 \\ V_{fc} = E_N - (R_m + R_F) I_0 \end{cases} \tag{3.3}$$

If the amplitude or the angular frequency is equal to zero, the output voltage of a fuel cell and that of the stack are presented as Eq. 3.3 and the voltage over the double-layer capacitor keeps constant with no charge or discharge. The output voltage and the resistance are corresponding to the stable voltage and the absolute value of slope of the polarization curve under current conditions.

3.3.1 *Feasibility Analysis for the Specialized DC/DC Converter*

As mentioned in Sect. 3.2, the specialized DC/DC converter for the fuel cell stack is composed of two DC/DC converters, as DC1 and DC2. In this part, DC1 and DC2 are both a non-isolated boost converter and the topology of them is presented (Fig. 3.5).

DC1 is comprised of the inductance L_1, the diode D_1, the IGBT-type semiconductor G_1 and the capacitor C_1. Similarly, DC2 is comprised of the inductance L_2, the diode D_2, the IGBT-type semiconductor G_2 and the capacitor C_2. G_1 and G_2 are controlled in the PMW mode and the frequency of PWM is the same while the duty cycle of PWM for them is D_{t1} and D_{t2} respectively. The variable u_{in}, i_{in}, u_o and i_o are the input voltage, input current, output voltage and output current of the specialized DC/DC converter. The variable i_{in1} and i_{o1} are the input current and output current

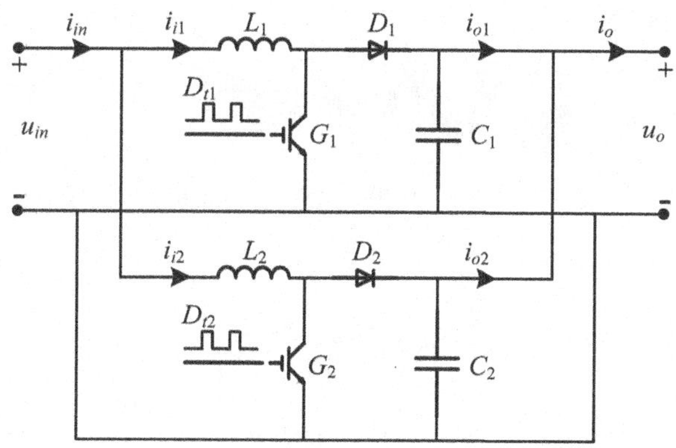

Fig. 3.5 Circuit of the specialized DC/DC converter

of DC1 and the variable i_{in2} and i_{o2} are the input current and output current of DC2. The input current of the boost converter must be no less than zero.

In the simulation model by Matlab/Simulink, the input of the specialized DC/DC converter is connected to equivalent circuit model of the fuel cell stack and the output of this converter is connected to a battery pack and a resistor R_{LD} as load. DC1 and DC2 are controlled to draw the target input current and the PID controller is used to complete their target by dynamically adjusting the duty cycle of G_1 and G_2. The target frequency for AC impedance is from 0.1 Hz to 1 kHz. If the frequency is lower than 0.1 Hz, the measured impedance would be dispersed for effective analysis. If the frequency is higher than 1 kHz, the inductance of the wire would dominate the measured impedance and it's not easy to be analyzed.

The parameters of the fuel cell stack and the converter is listed (Table 3.1) and no energy loss, resulting from the turning-on and turning-off of the semiconductor as well as the forward voltage of the diode, is considered due to complexity. Five frequency points of the current excitation like 0.1 Hz, 1 Hz, 10 Hz, 100 Hz and 1 kHz are simulated and the amplitude of the current excitation is 5 A. The dynamic output current and voltage of the fuel cell stack are shown (Figs. 3.6 and 3.7).

Table 3.1 Parameters of the fuel cell and the DC/DC converter

Parameter	Value	Parameter	Value
Inductance L_1	600 μH	PR R_{L1}	10 mΩ
Inductance L_2	600 μH	PR R_{L2}	10 mΩ
Capacitance C_1	600 μF	PR R_{C1}	10 mΩ
Capacitance C_2	600 μF	PR R_{C2}	10 mΩ
SF f_{switch}	20 kHz	NC N_{stack}	250
Nernst voltage E_N	1.2 V	DLC C_{dl}	2 F
Faradic resistance R_F	2 mΩ	MR R_m	1.2 mΩ

SF means the switching frequency. DLC means the double-layer capacitor. MR means the membrane resistance. NC means the number of cells. PR means the parasite resistance

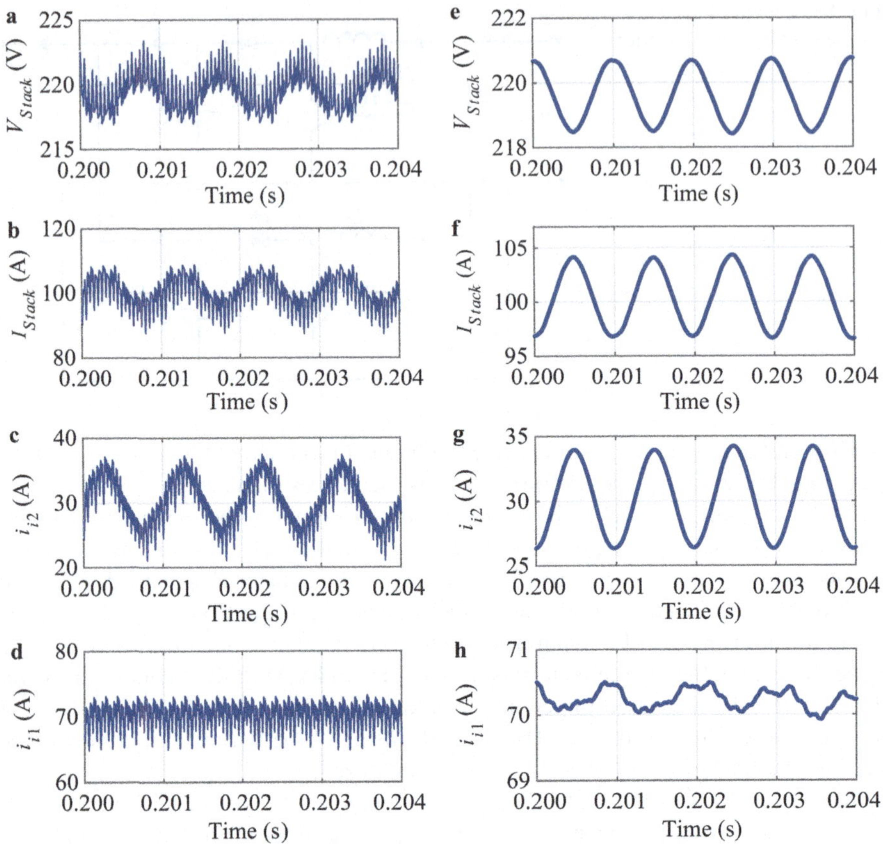

Fig. 3.6 **a** Original stack voltage at 1 kHz, **b** Original stack current at 1 kHz, **c** Original input current of DC1 at 1 kHz, **d** Original input current of DC2 at 1 kHz, **e** Filtered stack voltage at 1 kHz, **f** Filtered stack current at 1 kHz, **g** Filtered input current of DC1 at 1 kHz, **h** Filtered input current of DC2 at 1 kHz

The minimal output current of the specialized DC/DC converter is approximate to the input voltage of the converter divided by the resistor R_{LD}. The input current of DC2 is sum of a stable current and the current excitation and this applies to the second topology in Sect. 3.2. The target stable input current of DC1 is 70 A and that of DC2 is 30 A. Once the DC2 is activated, the stable input current of DC1 and that of DC2 are controlled coordinately to guarantee the stable current at 100 A of the fuel cell stack.

The impedance is calculated based on the input current and voltage of this converter by FFT technique. It demonstrates that the impedance at target frequency is equal to that of the theoretical value based on the equivalent circuit model. Furthermore, it verifies that the specialized DC/DC converter can inject current excitation signal to the fuel cell stack with wide frequency range and target amplitude.

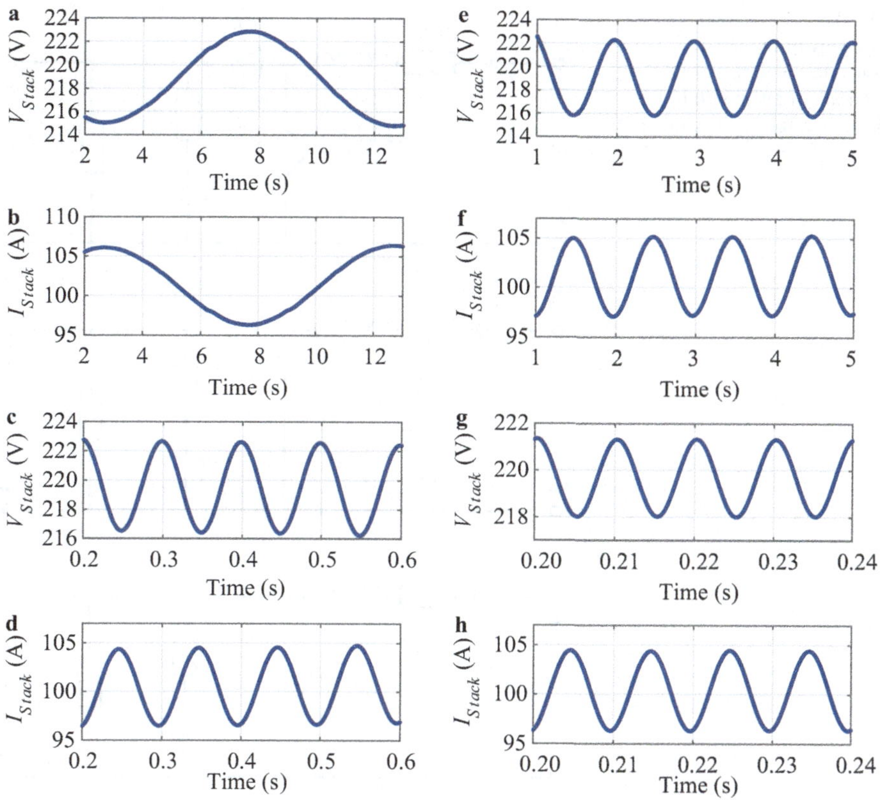

Fig. 3.7 **a** Filtered stack voltage at 0.1 Hz, **b** Filtered stack current at 0.1 Hz, **c** Filtered stack voltage at 1 Hz, **d** Filtered stack current at 1 Hz, **e** Filtered stack voltage at 10 Hz, **f** Filtered stack current at 10 Hz, **g** Filtered stack voltage at 100 Hz, **h** Filtered stack current at 100 Hz

3.3.2 *Feasibility Analysis for the Second Topology*

The second topology is also comprised of two DC/DC converters as DC1 and DC2 (Fig. 3.8a). DC1 is controlled to meet the requested power from load and DC2 is used to generate current excitation. The output of DC2 is connected to a resistor R_{LD}. DC2 is composed of the inductance L_1, the diode D_1, the IGBT-type semiconductor G_1 and the capacitor C_1. R_{L1} is the parasite resistor of L_1 and R_{C1} the parasite resistor of C_1. The duty cycle of PWM of G_1 is D_{t1}. The variable u_{in}, i_{in} and u_o are the input voltage, input current and output voltage of the converter respectively.

Turning-on and turning-off of the IGBT G_1 determines two different states of the converter DC2. When G_1 is turned on (Fig. 3.8b), the state equation for the circuit is as Eq. 3.4. When G_1 is turned off (Fig. 3.8c), the state equation for the circuit is as Eq. 3.5. In one period of PWM, the duration time during which G_1 keeps turned on is $D_{t1}T_{\text{switch}}$ and the duration time during which G_1 keeps turned off is $(1-D_{t1})T_{switch}$,

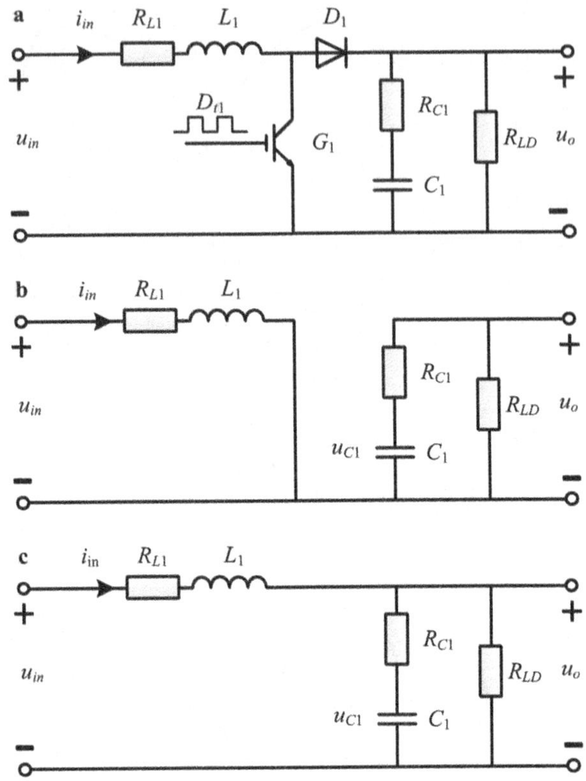

Fig. 3.8 The second topology of the DC/DC converter: **a** circuit of DC2, **b** circuit of DC2 with IGBT on, **c** circuit of DC2 with IGBT off

whereas T_{switch} is the period of PWM and it is inversely proportional to frequency of the PWM. The average model, combining Eqs. 3.4 and 3.5 together during the period, is presented as Eq. 3.6. The state variable is the voltage u_{C1} over the capacitor C_1 and the input current i_{in} and the input variable is the duty cycle D_{t1} and the output voltage u_o.

$$\begin{cases} u_{in} = i_{in}R_{L1} + L_1\frac{di_{in}}{dt} \\ C_1\frac{du_{C1}}{dt}(R_{C1} + R_{LD}) = -u_{C1} \\ u_o = -R_{LD}C_1\frac{du_{C1}}{dt} \end{cases} \tag{3.4}$$

$$\begin{cases} u_{in} = i_{in}R_{L1} + L_1\frac{di_{in}}{dt} + u_o \\ u_{C1} + R_{C1}C_1\frac{du_{C1}}{dt} = u_o \\ i_{in} - C_1\frac{du_{C1}}{dt} = \frac{u_o}{R_{LD}} \end{cases} \tag{3.5}$$

$$\begin{cases} u_{in} = i_{in}R_{L1} + L_1\frac{di_{in}}{dt} + (1 - D_{t1})u_o \\ u_{C1} + R_{C1}C_1\frac{du_{C1}}{dt} + D_{t1}R_{LD}C_1\frac{du_{C1}}{dt} = (1 - D_{t1})u_o \\ (1 - D_{t1})R_{LD}i_{in} = u_o + R_{LD}C_1\frac{du_{C1}}{dt} \end{cases} \tag{3.6}$$

$$\begin{cases} i_{in} = I_{in} + \hat{i}_{in}, u_o = U_o + \hat{u}_o \\ u_{C1} = U_{C1} + \hat{U}_{C1}, u_{in} = U_{in} + \hat{u}_{in} \\ d_{t1} = D_{t1} + \hat{d} \end{cases} \tag{3.7}$$

Based on the average model, the small signal model by imposing small fluctuation on the stable signal is presented as Eqs. 3.7 and 3.8, whereas I_{in}, U_o, U_{C1}, U_{in} and D_{t1} are the stable value for each variable respectively, and the second terms on right side of each equation are dynamic value for each variable respectively. To be noticed is that terms with higher order are neglected.

$$\begin{cases} L_1 \frac{d\hat{i}_{in}}{dt} = \hat{u}_{in} - R_{L1}\hat{i}_{in} - (1 - D_{t1})\hat{u}_o + U_o\hat{d} \\ R_{LD}C_1 \frac{d\hat{u}_o}{dt} = (1 - D_{t1})\hat{i}_{in} - I_{in}\hat{d} - \hat{u}_o \end{cases} \tag{3.8}$$

The relation, defined as the transfer function $G_{id}(s)$ between the fluctuation of duty cycle and that of the input current is derived as Eq. 3.9. If an ideal voltage source is connected to the input of DC2 instead of the fuel cell stack, the fluctuation of input voltage is equal to zero, the transfer function can be simplified as Eq. 3.10. If the fuel cell stack is connected to the input of DC2 and resistance of the stack is R_{in}, the transfer function is derived as Eq. 3.11.

$$\begin{cases} f(\cdot) = (R_{LD}C_1 s + 2)U_o\hat{d}(s) \\ \hat{i}_{in}(s) = \frac{(R_{LD}C_1 s+1)\left(\hat{u}_{in}(s) - R_{L1}\hat{i}_{in}(s)\right) + f(\cdot)}{R_{LD}L_1C_1 s^2 + L_1 s + R_{LD}(1-D_{t1})^2} \end{cases} \tag{3.9}$$

$$\begin{cases} f(\cdot) = R_{LD}(1 - D_{t1})^2 + R_{L1} \\ G_{id}(s) = \frac{\hat{i}_{in}(s)}{\hat{d}(s)} = \frac{(R_{LD}C_1 s+2)U_o}{R_{LD}L_1C_1 s^2 + (L_1 + R_{LD}C_1R_{L1})s + f(\cdot)} \end{cases} \tag{3.10}$$

$$\begin{cases} f(\cdot) = R_{LD}(1 - D_{t1})^2 + R_{L1} + R_{in} \\ G_{id}(s) = \frac{(R_{LD}C_1 s+2)U_o}{R_{LD}L_1C_1 s^2 + (L_1 + R_{LD}C_1(R_{L1} + R_{in}))s + f(\cdot)} \end{cases} \tag{3.11}$$

Based on Eq. 3.11, the tendency of the transfer function with angular frequency is derived as Eq. 3.12. Obviously, the gain of the transfer function is limited through adjusting amplitude of the duty cycle fluctuation when the angular frequency is very high. That is to say, the DC/DC converter is not suitable for AC impedance at such high frequency and the inductance L_1 is the dominating factor.

$$G_{id}(j\omega) \approx \begin{cases} \frac{2U_o}{R_{LD}(1-D_{t1})^2 + R_{L1} + R_{in}}, if \omega \to 0 \\ 0, if \omega \to \infty \end{cases} \tag{3.12}$$

The derivation of the transfer function gain with respect to frequency is presented as Eq. 3.13. Parasite resistors of the capacitor and the inductance is significantly lower than R_{LD} as load, so they are just neglected for this part. Then the critical frequency at the maximal of the transfer function gain is derived as Eq. 3.14 and the maximal is as Eq. 3.15.

$$\begin{cases} \omega = j2\pi f \\ \frac{d|G_{id}(j2\pi f)|}{df} = 0 \end{cases} \tag{3.13}$$

$$f_{cr} = \frac{1}{2\pi}\sqrt{\frac{R_{LD}(1-D_{t1})^2 + R_{L1} + R_{in}}{R_{LD}L_1C_1}} \approx \frac{1-D_{t1}}{2\pi\sqrt{L_1C_1}} \tag{3.14}$$

$$\begin{aligned} |G_{id}(j2\pi f_{cr})| &= \frac{U_o\sqrt{4+(2\pi f_{cr}R_{LD}C_1)^2}}{2\pi f_{cr}[L_1 + R_{LD}C_1(R_{L1}+R_{in})]} \\ &\approx \frac{U_o\sqrt{4+(2\pi f_{cr}R_{LD}C_1)^2}}{2\pi f_{cr}L_1} \end{aligned} \tag{3.15}$$

In comparison with the resonance frequency of the inductance and the capacitor is serial, the IGBT-type semiconductor G_1 and the diode D_1 together change the resonance frequency which is multiplied by the non-effective duty cycle (1-D_{t1}) based Eq. 3.14. The reason is that only when the semiconductor G_1 is turned off, the inductance L_1 and the capacitor C_1 are connected.

According to Eq. 3.15, when the stable output voltage of DC2 is increased, the transfer function gain is increased proportionally no matter how large the critical frequency, the inductance and the load resistance are. If the amplitude of the duty cycle fluctuation is increased, the amplitude of the current excitation as well as the dynamic change of the input current would be significantly enlarged around the resonance frequency and the overcurrent protection may be activated which is not wanted in actual application.

Based on Eqs. 3.12 and 3.15, the relation between the transfer function gain and the frequency is plotted (Fig. 3.9). In order to guarantee an appropriate amplitude of current excitation, the target frequency for AC impedance is suggested to be close to the critical frequency but the amplitude should be controllable for stability and safety. When the DC/DC converter is designed to integrate the online AC impedance

Fig. 3.9 Gain of transfer function and frequency

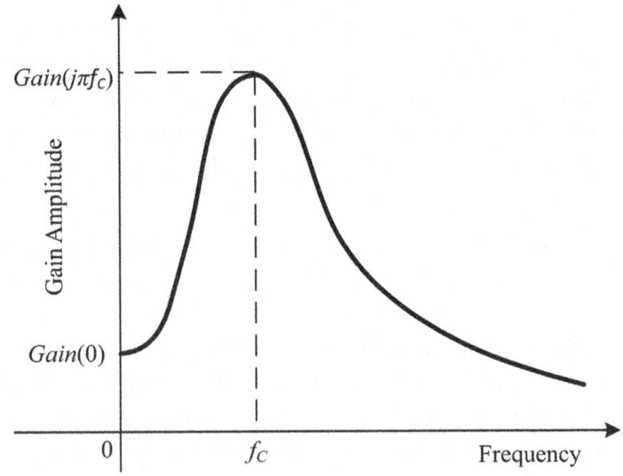

Table 3.2 Parameters of the fuel cell and the DC/DC converter of the second topology

Parameter	Value	Parameter	Value
Inductance L_1	1 mH	PR R_{L1}	1 mΩ
Capacitance C_1	300 μF	PR R_{C1}	1 mΩ
Stack Resistance R_{S1}	10 mΩ	Stack Resistance R_{S2}	10 mΩ
Stack Capacitance C_{S1}	0.4 F	Contact Resistance S_1 & S_2	1 mΩ
Load Resistance R_{LD}	20 Ω	Pre-charge Resistance R_1	20 Ω
Switching Freq	10 kHz	On-Resistance D_1 and G_1	1 mΩ

PR means the parasite resistance. Switching Freq. means the switching frequency of the IGBT-type semiconductor

measurement function, the selection of the frequency for AC impedance and that of the critical frequency are very important work.

For simplification, an electronic load in constant current mode is selected as DC1 and the model is set up by Matlab/Simulink. The parameters for the equivalent circuit model of the stack and components of the converter are listed (Table 3.2). The simulation result for the current excitation signal and the voltage response of the stack at 10 Hz and 100 Hz is illustrated (Fig. 3.10).

By substituting the component parameters of this converter into Eq. 3.9, the theoretical transfer function gain is obtained. Then the theoretical gain can be compared with that of the simulation result (Table 3.3). The error is defined as percentage of the difference between the theoretical gain and the simulation result divided by the theoretical gain. The error is within 7% for frequency lower than 500 Hz. At 1 kHz, the error reaches 10% and the reason is that the switching frequency of the semiconductor G_1 is just 10 kHz and to generate the current excitation at 1 kHz, the effective controllable duty cycle in one period is only 10 points and it's full of uncertainty. At 100 Hz and 200 Hz, the error reaches 7% because the frequency is around the critical frequency and the theoretical gain is significantly larger than others. In this case, if we take a look at the overall input current of DC2, the linearity assumption of the small signal model of DC2 won't be sufficiently satisfied as Eqs. (3.3)–(3.8).

Even though the error of the current excitation gain with respect to frequency exists for the converter DC2, the calculated AC impedance based on the current excitation and the voltage response shows are well matched with the theoretical resistance of the fuel cell stack, using the FFC technique.

In a word, the small signal model of the DC/DC converter can quantify its dynamic response accurately and on this basis, the generation of current excitation can be realized by fluctuation of the duty cycle. This method can be applied to situations where the topology of the converter is different and the small signal model provides theoretical guidance for the converter design for online AC impedance measurement.

If frequency of the duty cycle fluctuation is varied, the length of the vector line OM and that of the vector line OS are changed. If amplitude of the duty cycle fluctuation is varied, the length of the vector lien OW and that of the vector line OX are changed. Both cases will result in variation of the equivalent impedance OK. The length of the vector line OM increases in proportional to the frequency but the length of the vector line OS decreased significantly with the increased frequency. In this case, the increased frequency can result in the equivalent impedance lying in the first phase of the Nyquist plot.

3.4 Weak Signal Extraction at High Speed and High Precision

The output power of the fuel cell system for powertrain in vehicle usually reaches several tens of kilo watts and the fuel cell stack is composed of several hundreds of cells in serial to lift the voltage. For example, the output current of the stack is 500 A with the current density at 2 A/cm^2 and the output voltage of the stack ranges from 220 to 350 V for MIRAI developed by Toyota Motor Corporation. The amplitude of the sinusoidal current excitation is recommended to be 1% to 10% of the steady output current of the stack. When the output current flows through a shunt resistor, the voltage difference over it provides a good way to measure the current with high precision and good ability of heat dissipation. For example, the nominal value of a shunt resistor is 75 mV@500A and the measured voltage of the current excitation over the shunt resistor is less than 5 mV. The voltage response of the fuel cell in normal operating state is also less than 5 mV with the high frequency resistance lower than 0.5 mΩ. The voltage response of the fuel cell stack is approximately sum of that of each cell.

After the DC/DC converter is commanded to superimpose the current excitation to the fuel cell stack, to calculate the AC impedance exactly, it is required that the voltage response of the stack and the current excitation are measured simultaneously and the voltage response of each fuel cell and the current excitation are measured simultaneously.

Something important should be emphasized which is encountered and overcome in our research. To start with, the conventional hall-effect current sensor usually outputs voltage at high current to voltage ratio and the guaranteed accuracy is around 0.5% at most. When it is used to measure the current excitation and the voltage response of a fuel cell, the target weak signal is immersed in the noise and the signal noise ratio is very low even though the frequency band of the sensor is wider than that of the target AC impedance. In addition, the amplitude attenuation and the phase delay of the hardware-based signal extraction circuit ought to be compensated in advance and the calibration is necessary because impedance is the voltage divided by current and the error is enlarged in this way. Furthermore, to measure impedance of all cells needs more signal extraction circuits but the design of the CVM (cell voltage monitor) for the fuel cell system is limited by cost and volume. Finding solution to sharing one suite of signal extraction circuit for more cells is meaningful and challenging. Finally, fuel cells are connected in serial and the common-mode voltage at the positive end and the negative end of a cell is very high, so the common-mode voltage should be eliminated to accurately measure the cell voltage.

Our solution is presented (Fig. 3.12) for these problems, including a shunt resistor, the hardware-based signal processing circuit for the stack voltage, that for the output current of the stack, that for the voltage of each fuel cell, the method for sharing these circuits and other circuits for the microprocessor. All these circuits are integrated into the CVM.

The method for sharing these signal processing circuits are illustrated (Fig. 3.13). It is realized by setting the trigger command in order to activate the target cell number and for example, the 4–16 wired decoder and the optical coupler equipment can be combined and the microprocessor sends the trigger signal to the decoder.

The circuit for weak signal extraction is presented (Fig. 3.14). The high common-mode differential circuit is used to eliminate the high common-mode voltage over each cell. Different signal processing circuits are proposed for the steady voltage

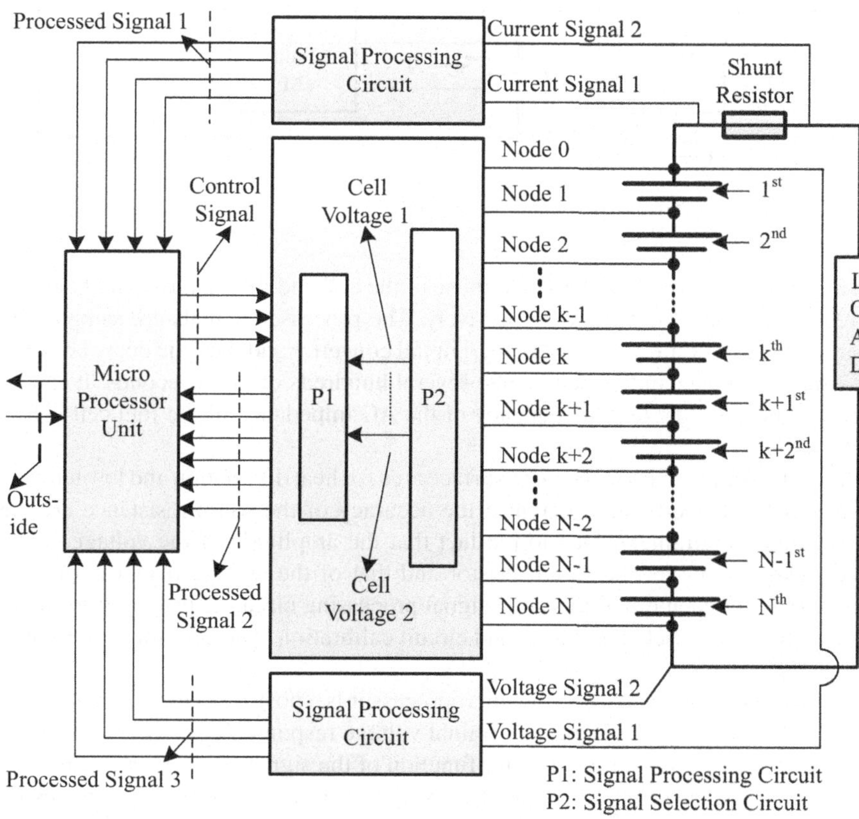

Fig. 3.12 Scheme of the device for weak signal extraction

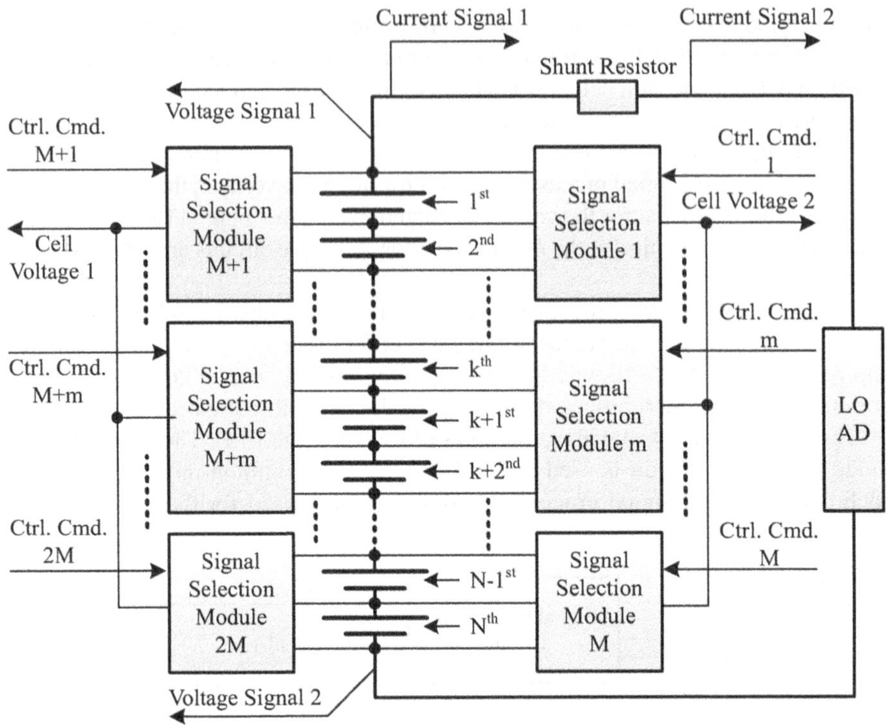

Fig. 3.13 Scheme of cell selection

and the voltage response of the stack, those of the cell and the steady current and the current excitation of the stack, respectively. The processed signals are sampled by the microprocessor through the Analog/Digital converter and the time delay between the sampled voltage and current is just several hundreds of microseconds. It can be neglected because the target frequency of the AC impedance of the fuel cell is less than 1 kHz.

About the shunt resistor, the large surface area for heat dissipation and low temperature sensitivity coefficient guarantee the accuracy of the shunt resistance are the fundamental requirement. Due to the fact that the amplitude of the voltage of the current excitation over the shunt resistor and that of the voltage response of each cell are close to each other, the same signal processing circuit is proposed for them and it simplifies the scheme design and circuit calibration. The solution is presented (Fig. 3.15).

The sampled voltage U_{sV} by the microprocessor is a post-processed signal, by the hardware circuit (Fig. 3.15), of the original voltage response U_{origV} of a fuel cell to the current excitation, and the transfer function of the signal processing circuit is as H_{spc1}. Similarly, the sampled voltage U_{sC} by the microprocessor is a post-processed signal, by the hardware circuit (Fig. 3.15), of the converted voltage over the shunt resistor R_{SHR} of the current excitation I_{origC} to the fuel cell, and the transfer function

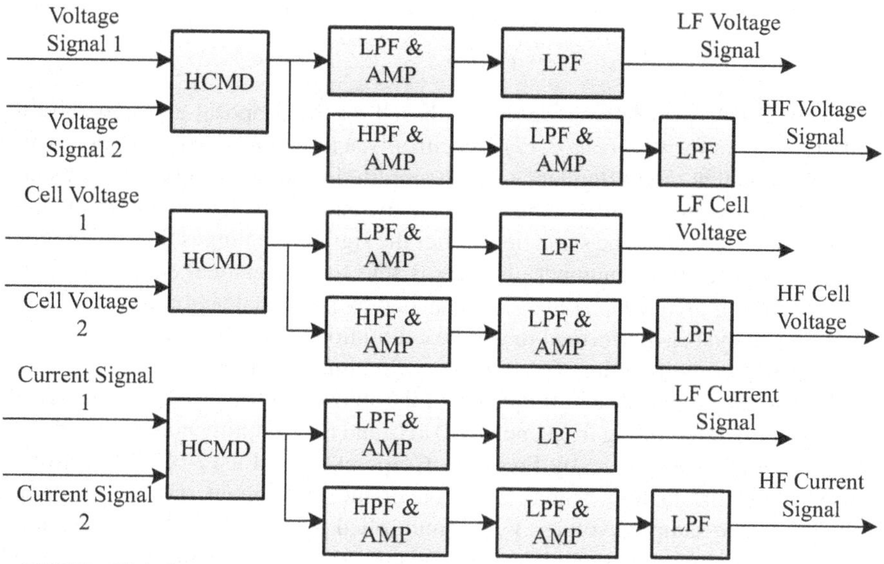

HCMD: High Common-Mode Differentiator
LPF: Low-Pass Filter **LF**: Low Frequency
HF: High Frequency **HPF**: High-Pass Filter
AMP: Amplitude Attenuation and Offset Voltage Adjustment

Fig. 3.14 Scheme of signal processing

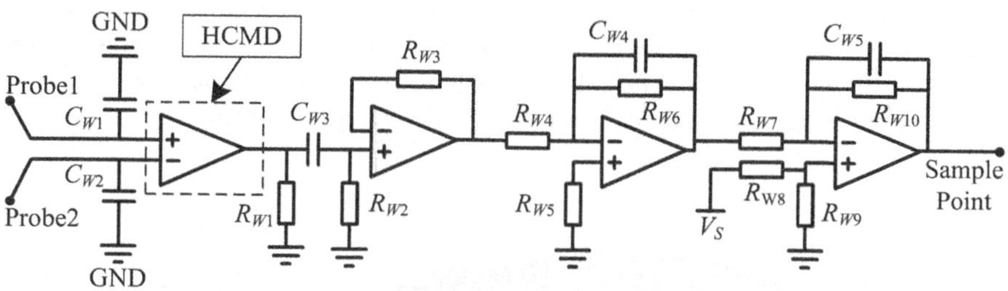

Fig. 3.15 An example of the signal processing circuit

of the signal processing circuit is as H_{spc2}. The impedance Z_{fc} of the fuel cell is calculated as Eq. 3.16 and if the transfer function H_{spc1} and H_{spc2} are designed to be the same, the calculation of impedance is greatly simplified. However, the actual state of the circuit component always deviates from its nominal value and the calibration is required to acquire the difference of the amplitude attenuation and that of the phase delay between the transfer function H_{spc1} and H_{spc2}, based on the Eq. 3.16. That is to say, we don't need to do calibration for each circuit and it's a great contribution. The impedance calculation of the stack is in the same way as well as the circuit calibration and it won't be derived here.

$$Z_{fc} = \frac{U_{origV}}{I_{origC}} = \frac{U_{sV}/H_{spc1}}{U_{sC}/H_{spc2}} \times R_{SHR} = \frac{U_{sV}}{U_{sC}} \times R_{SHR} \tag{3.16}$$

A new CVM (Fig. 3.16) is developed based on the proposal and a method for calibration is put forward (Fig. 3.17). Specifically, a function signal generator, which generates signal at target frequency, target amplitude and target offset, is connected to signal processing circuits of the CVM for the current excitation and the voltage response of a fuel cell at the same time. After the signal sampling is finished at preset frequency and for target number, the data is sent to the personal computer via the CAN (controller area network) bus and decoded by the signal analyzer VN1630 and the offline analysis is performed to get the calibration result.

For example, the function signal generator outputs a voltage of which the offset is 0.7 V and the amplitude and frequency of the sinusoidal wave is 2 mV and 320 Hz respectively. The sampling frequency is 10 kHz and the sampling number 1,000. The microprocessor is MC9S12 by Freescale Corporation and a 12-bit A/D converter is used to obtain analogy signal. A group of the experiment result is presented (Fig. 3.18). The sampled voltage V_{sMCU} multiplied by 5/4,095 is the voltage at the input of the microprocessor and the sampled current I_{sMCU} in multiplied by 5/4,095 is the converted voltage at the input at the microprocessor. Looking back on the Eq. 3.16, the factor 5/4095 can be neglected because it exists in the numerator and the denominator, so using the sampled voltage V_{sMCU} and the sampled current I_{sMCU} instead is mathematically reasonable. It verifies that the CVM can extract the weak signal from the steady signal with the target cell selectable.

Besides, the signal processing circuit is shared by all fuel cells so the calibration of the signal processing circuit only needs to be done for one cell for the CVM. Six groups of experiment result are listed (Table 3.4) for the first cell with the same current excitation at 320 Hz. The amplitude attenuation and phase delay is calculated and the result validates the function of the CVM with high stability.

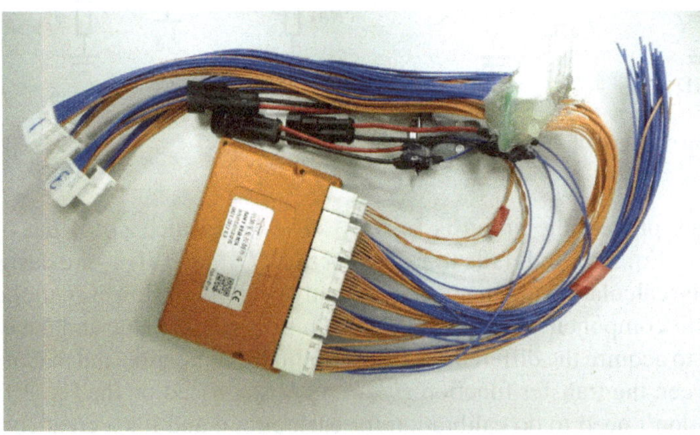

Fig. 3.16 The developed cell voltage monitor

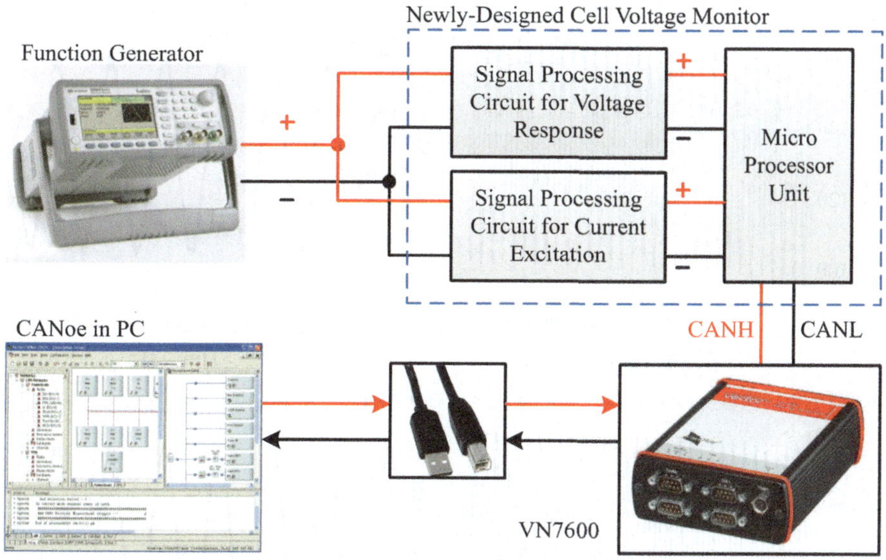

Fig. 3.17 Calibration method for the signal processing circuit

3.5 Experiment Based Analysis of Feasibility for System Application

Taking into consideration the giant difficulty in designing the integrated DC/DC converter in the first topology, a boost DC/DC converter (Fig. 3.19), is developed for the component DC2 in the second topology and it is used to generate the current excitation. The component parameter of the converter is listed (Table 3.5). For high reliability, the converter is equipped with over current protection for both the input current and the output current, over voltage protection and under voltage protection for both the input voltage and the output voltage, short circuit protection and load dump protection. A pre-charge circuit is designed for the capacitor in parallel with the input of the converter. Based on principle of the second topology as Sect. 3.3.2, efforts will be made to realize the system function in this converter with just the open-loop control.

A lithium-ion battery pack is our test object instead of the fuel cell stack for reasons. Firstly, by selectively arranging the battery in serial or in parallel or in a combined way, the electrochemical impedance spectroscopy of the lithium-ion battery is similar to that of a fuel cell at the concerned frequency range. Secondly, operating the fuel cell stack of high voltage with a test plant is very complicated and costive but operating the battery pack is more convenient. Finally, safety is our first priority and uncertain risk exists in our experiment but the battery pack is more safe with short-circuit protection and battery management system. The nominal capacity of the lithium manganese battery pack is 170 Ah and it is arranged as 5 batteries in

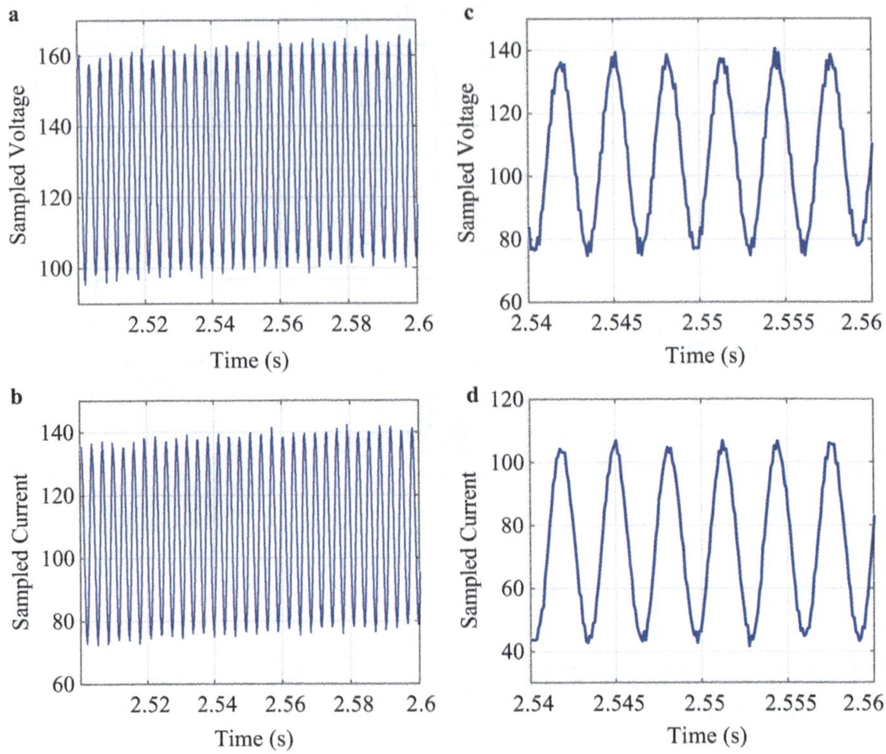

Fig. 3.18 **a** Sampled voltage at 320 Hz for the 01st cell, **b** Sampled current at 320 Hz for the 01st cell, **c** Sampled voltage at 320 Hz for the 15th cell, **d** Sampled current at 320 Hz for the 15th cell

Table 3.4 Calibration result of the signal processing circuit at 320 Hz

Voltage Amp	Voltage phase	Current Amp	Current phase	Amp. difference	Phase difference
207.92	17.23°	211.77	26.808°	1.019	9.58°
207.85	185.82°	211.25	195.425°	1.016	9.60°
206.93	195.91°	211.45	205.452°	1.022	9.54°
207.58	10.87°	211.27	20.413°	1.018	9.55°
207.69	201.17°	211.42	210.825°	1.018	9.66°
206.72	197.86°	210.90	207.550°	1.020	9.69°

Voltage Amp. and phase means the FFT amplitude and phase of the sampled voltage. Current Amp. and phase means the FFT amplitude and phase of the sampled current. Amp. Difference is the current amplitude divided by the voltage amplitude. Phase difference is the current phase minus the voltage phase.

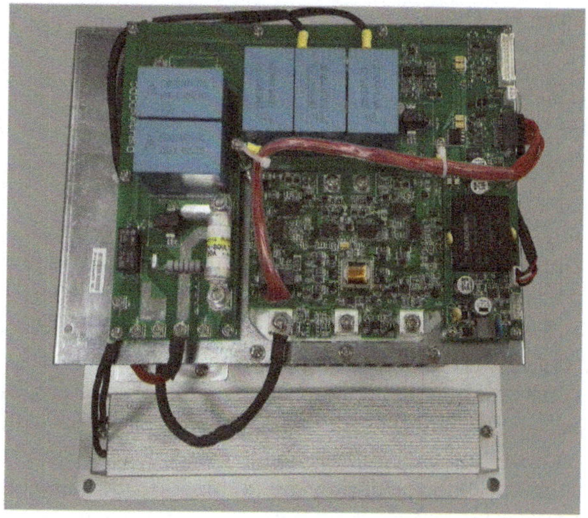

Fig. 3.19 The developed DC/DC converter for experiment

Table 3.5 Parameters of the developed DC/DC converter

Parameter	Value	Parameter	Value
Input Capacitance	200 μF	Input Voltage	80–150 V
Output Capacitance	300 μF	Input Current	0–60 A
Inductance	1 mH	Output Voltage	150 –300 V
Switching Frequency	10 kHz	Output Current	0–30 A
Amplitude of CE	0–15 A	Freq. of CE	0.1 Hz –1 kHz

CE means the current excitation

serial at first and 24 such connected batteries in parallel, provided by the Key Power Corporation.

A software is designed to control the DC/DC converter and monitor state of the converter and the battery pack using the NI LabVIEW tool. The experiment data is logged by the software and it facilitates the offline data analysis.

In addition, the nominal value of the shunt resistor is 75 mV/300 A at 25 °C and the calibration experiment shows that the shunt resistor is 0.2565 mΩ (Table 3.6).

The test plant is set up as Fig. 3.4. To be specific, the battery test system (BTS600) by Digatron Power Electronics Corporation draws a constant current from the battery and the electronic load (IT8838H) by ITECHS corporation works in constant resistance mode and it is connected to the output of the converter. In our experiment, the BTS600 draws current at 70 A, the converter draws 30 A and the resistance of IT8838H is 20 Ω. Then the DC/DC converter is controlled to generate current excitation on its input current at different frequency and amplitude. The total input

Table 3.6 Calibration of the shunt resistor

Current (A)	Voltage (mV)	Current (A)	Voltage (Mv)
5.44	1.3	16.87	4.4
6.98	1.8	26.81	6.8
8.92	2.3	34.24	8.8
11.71	3.1	/	/

current of the BTS600 and the converter is the output current of the battery pack so the current excitation is superimposed on the battery pack.

In the first experiment, the wave of the sampled current excitation is illustrated (Fig. 3.20) and the target frequency is 1 Hz, 10 Hz, 100 Hz and 1 kHz. The time constant of the dynamic response of the signal processing circuit is relatively long, because the cutoff frequency of the high-pass filter in the circuit is very low. The influence of the time constant on the current excitation is the gradually increased steady sampling voltage during the continuous sampling. It is verified that the designed

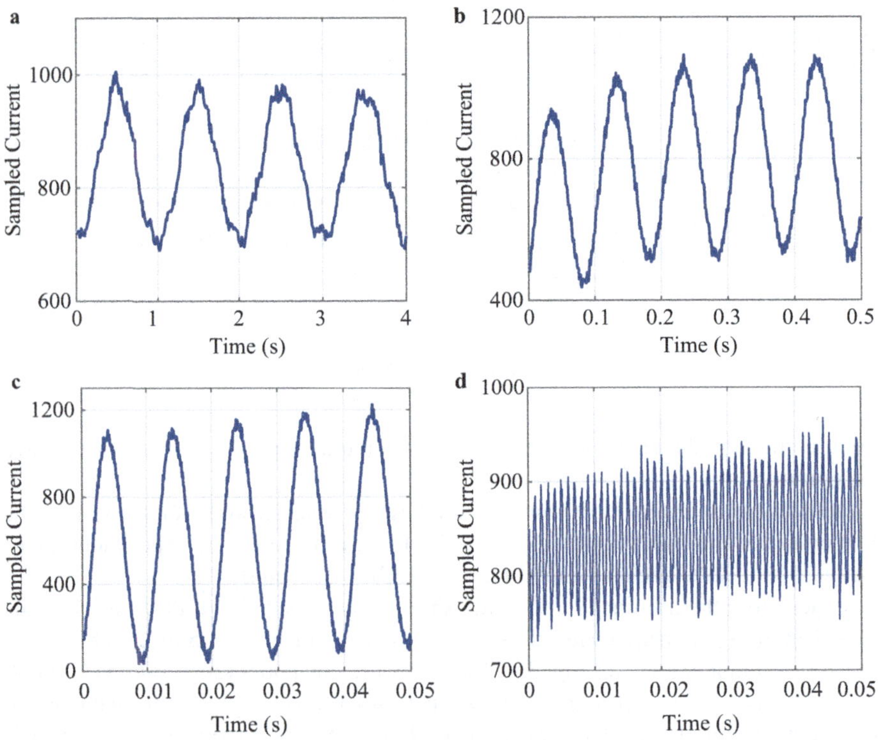

Fig. 3.20 **a** Sampled current at 1 Hz, **b** Sampled current at 10 Hz, **c** Sampled current at 100 Hz, **d** Sam-pled current at 1 kHz

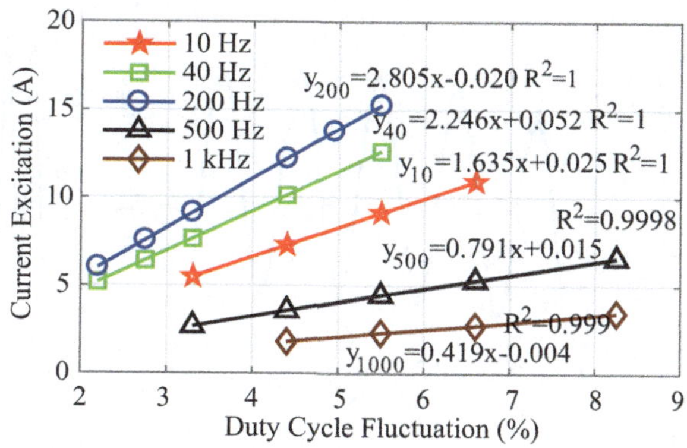

Fig. 3.21 Amplitude of current excitation and the duty cycle fluctuation at different frequency

DC/DC converter can inject current excitation to the battery pack and the CVM can extract the current excitation signal.

In the second experiment, the varied duty cycle excitation to the DC/DC converter at a frequency is performed. The test result is shown (Fig. 3.21). The first-order linear fitting equation for frequency, at 10, 40, 200, 500 Hz and 1 kHz, and the goodness of curve fitting is presented as well. It is verified that the relation between the amplitude of the duty cycle excitation as studied in Sect. 3.3.2 and that of the current excitation is highly linearized and the amplitude of the current excitation increases linearly with the increased amplitude of the duty cycle excitation. To be noticed is that the intercept of the fitting equation at each frequency is less than 0.1 A and this may result from the ripple current and noise at the crest and trough of the sampled wave as well as the dynamic response of the circuit (Fig. 3.17). Even so, it doesn't affect the target relation with relatively low amplitude.

The slope of the fitting equation is just the measured gain function as Eq. 3.11 and The comparison between the gain from experiment and the theoretical gain from component parameters is made (Fig. 3.22). Obviously, they match well with

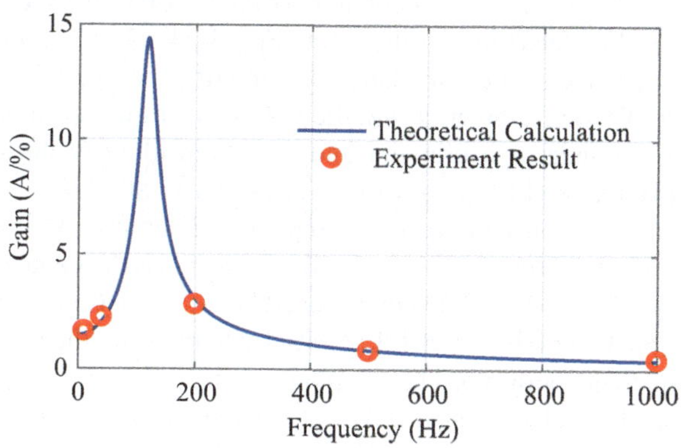

Fig. 3.22 Comparison of theoretical gain and experiment data

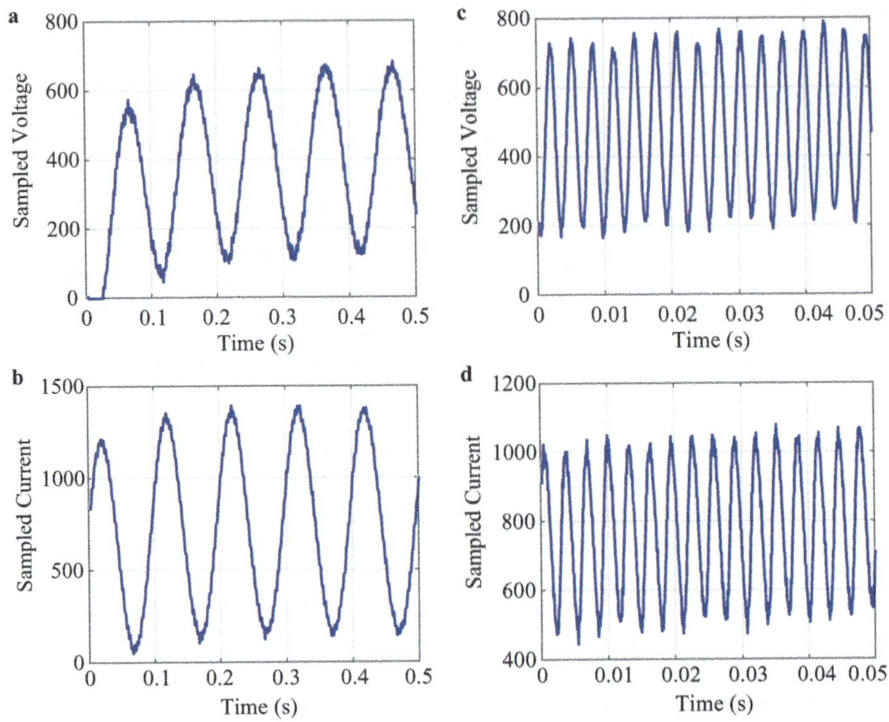

Fig. 3.23 **a** Sampled voltage at 10 Hz for the 16th cell, **b** Sampled current at 10 Hz for the 16th cell, **c** Sampled voltage at 320 Hz for the 8th cell, **d** Sampled current at 320 Hz for the 8th cell

each other and it demonstrates that the small-signal model can exactly quantify the dynamic response of the DC/DC converter to the duty cycle excitation.

Furthermore, in the third experiment, the current excitation and the voltage response of two batteries are simultaneously measured by the CVM (Fig. 3.23). The impedance is calculated based on FFT technique for four batteries and they are illustrated in Nyquist plot (Fig. 3.24). The higher the frequency, the lower the amplitude of the impedance. For the time being, the experiment result only tells that the DC/DC converter and the CVM function well together to realize the online AC impedance measurement and they help identify the uniformity between batteries. The realized frequency range is from 1 to 800 Hz.

Based on conclusion of the former three experiment, the fourth experiment is performed on a 30 kW fuel cell system by Hydrogenics Corporation from Canada and the 60 kW (Fig. 3.25) interleaved boost DC/DC converter by Sinohytec Corporation from China (Fig. 3.26). The target is to measure the HFR of the fuel cell stack with the stack in operation. The test plant for the AC impedance is set up as Fig. 3.4. The stack has 120 cells in serial and the supply subsystem of air and hydrogen is realized by the former corporation and the cooling subsystem is integrated by the latter corporation. The fuel cell system purges water from the anode chamber periodically and this varies the water content in the stack periodically. Only the voltage of two adjacent cells are available.

The research result by Toyota Motor Corporation shows that the impedance at frequency ranging from 300 Hz to 1 kHz is very close and taking the difficulty in application to the DC/DC converter into consideration, the impedance at 300 Hz is chosen as the HFR. This is just our reference and no more data is given for the fuel cell system by the supplier, so in this experiment we decide to measure the HFR at 320 Hz.

The room temperature is 35 °C and the relative humidity of air is 40%. One piece of the sampled voltage response of a cell and the sampled current excitation is presented as Fig. 3.27 and the sampling frequency is 10 kHz.

In comparison with the wave (Fig. 3.23), the sampled voltage response and the sampled current excitation are distorted from the sinusoidal signal even though the duty cycle excitation of the converter (Fig. 3.18) is the same. This originates from difference between the BTS600 and the 60 kW DC/DC converter. To be specific, the BTS600 guarantees that the drawn current is constant with little fluctuation even though the output voltage of the battery pack varies with the current excitation. On the contrary, the 60 kW DC/DC converter acquires the dynamic output voltage and

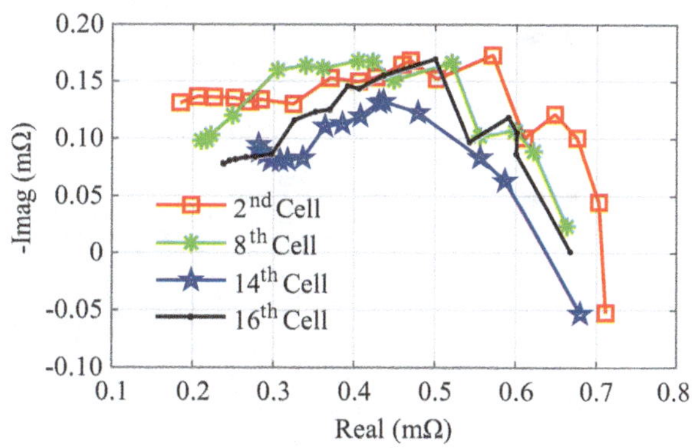

Fig. 3.24 Impedance spectroscopy of four battery cells

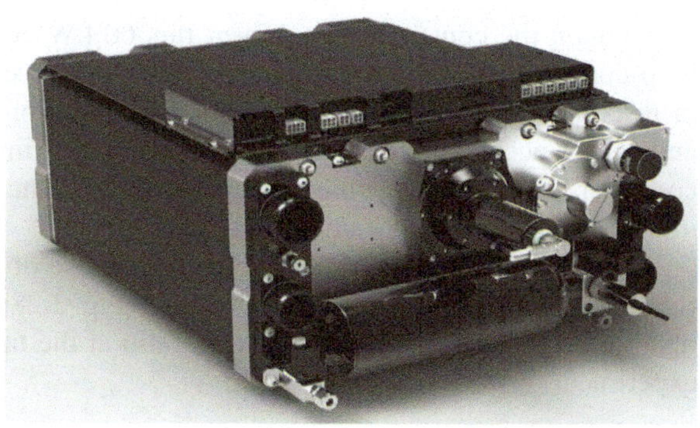

Fig. 3.25 HD30 by hydrogenics corporation

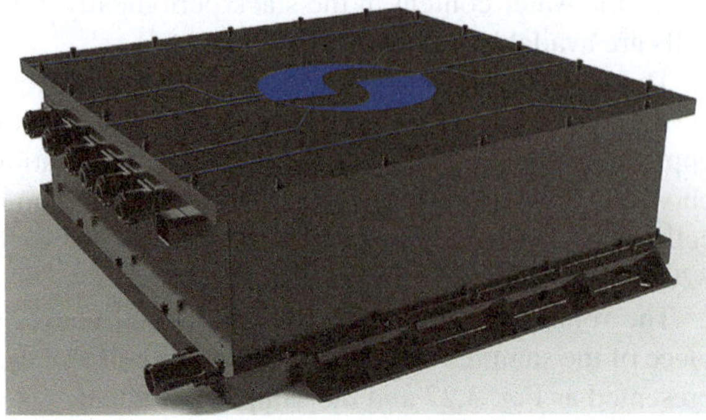

Fig. 3.26 The 60 kW DC/DC converter by sinohytec corporation

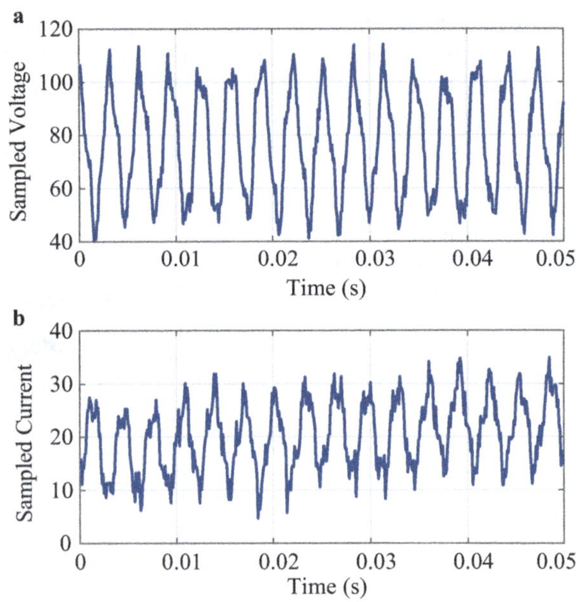

Fig. 3.27 **a** Sampled voltage of the fuel cell in stack, **b** Sampled current of the fuel cell stack

to maintain the constant input power, this 60 kW converter is operated in closed-loop control mode. That is to say, the 60 kW converter changes its input current in response to the input voltage with slower response and uncertainty like noise and sensor accuracy. The current excitation superimposed on input current of the converter (Fig. 3.18) plus the current response of the 60 kW converter is the result in the fourth experiment.

The spectrum of the sampled voltage response of two cells together and that of the sampled current excitation is calculated (Fig. 3.28) based on FFT technique. The distorted signal is of high signal noise ratio at the target frequency 320 Hz so the experiment result is reliable.

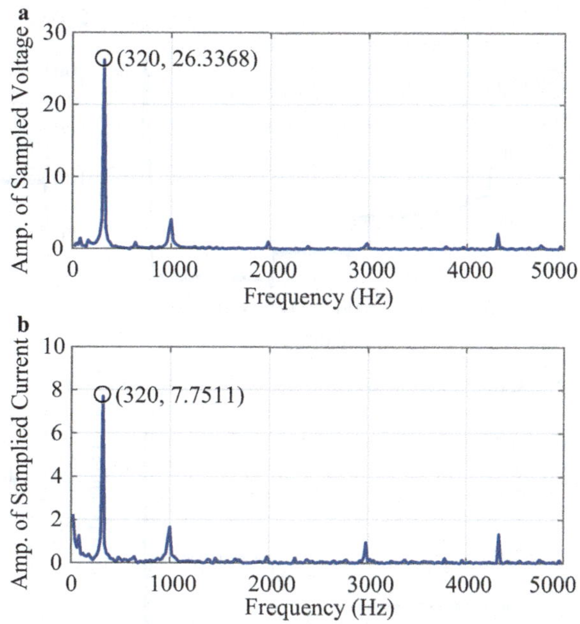

Fig. 3.28 **a** FFT analysis of the sampled voltage, **b** FFT analysis of the sampled current

We plot the HFR at 320 Hz (Fig. 3.29) and the operating conditions of the stack (Fig. 3.30), including the output voltage, the output current, the air flux and the coolant temperature at the stack cooling inlet. The fuel cell stack operates with output current as 75 A for some time and then the converter (Fig. 3.18) increases its steady input current as 25 A through the test. The output current of the stack maintains for some time at 100, 150 and 200 A in sequential until the test ends. When the operating condition of the stack at each current is steady, the current excitation injection to the stack is activated every 1 min and every time the activation lasts until the CVM has sampled signals for two complete group.

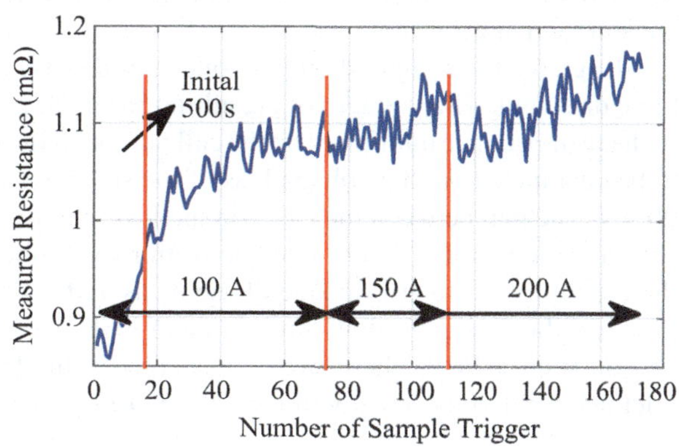

Fig. 3.29 Resistance of 2 cells at 320 Hz with time

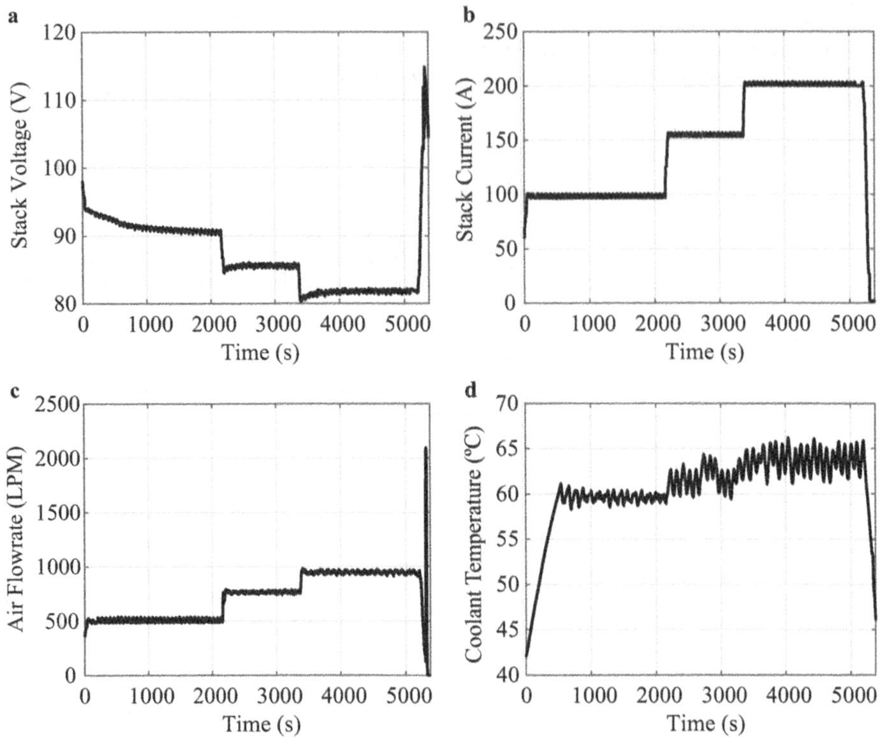

Fig. 3.30 **a** Stack voltage, **b** Stack current, **c** Air flow rate, d Coolant temperature at stack inlet

When the stack output current maintains at 100 A, the HFR of the measured two cells increase quickly at first and then it tends to be stable. These two cells lie in the end far away from the inlet manifold of the air supply and the hydrogen supply, as Fig. 3.30. From 0 to 500 s, the coolant temperature at the stack cooling inlet increases from 42 to 60 °C. From 500 to 2,000 s, the HFR increases from 0.942 mΩ to 1.113 mΩ and it tells that the water content in the stack decreases gradually with the stable coolant temperature.

The gradually decreased output voltage of the stack shows that the fuel cell tends to be dried out. The voltage loss resulting from the increase of the HFR is 1.026 V with the impedance multiplied by the cell number if the uniformity of cells is assumed to be guaranteed but the voltage loss of the stack is 1.6 V decreased from 92.5 V to 90.9 V. The gap between the two voltage loss may result from the nonuniformity of cells and the possible deteriorated activation overvoltage in the MEA. For example, the capability of water drainage of the measured cells is the lowest because the air flow and the hydrogen flow through them is the worst resulting from the structure design. If the HFR of these two cells increases, the HFR of other cells close to the inlet manifold of the stack will increase more significantly.

When the stack output current maintains at 150 and 200 A, the output voltage of the stack shows no significant trend of variation with time but the HFR increases with visible speed. It means that the fuel cell is being dried out as well but the activation overvoltage is improved to some extent. Besides, we can see that the measured HFR fluctuates for each constant current and it may result from the control of purge valve at the anode outlet to purge accumulated water and nitrogen in the anode chamber. The action of the purge valve is unknown.

Next, the experiment result will be analyzed in a qualitative way with respect to the capability of water drainage. The relative humidity RH_{atm} of air at the inlet manifold of the stack, and the partial pressure $P_{W,Ca,Out}$ of water vapor at the outlet manifold of the stack where waste air is exhausted, are calculated as Eq. 3.17, whereas $P_{Ca,Out}$ is the air pressure at the outlet manifold, λ the excessive stoichiometry ratio and α the net water flux per generated water, as byproduct of reaction, in the direction from the anode to the cathode. The variable α is unknown but it is sure that it ranges from −0.5 to zero. The reason is that the reaction in the anode catalyst layer doesn't generate water, there is no humidifier in the hydrogen supply system to import water and the purge valve is periodically opened to remove water, so the actual water transfer is from the cathode to the anode definitely. Based on Eq. 3.17, it is known that the partial pressure $P_{W,Ca,Out}$ of water vapor increases with the increased variable α.

$$\begin{cases} x_{W,Ca,In} = \frac{RH_{atm} \times P_{sat}(T_{atm})}{P_{atm}} \\ P_{W,Ca,Out} = P_{Ca,Out} \frac{\lambda x_{W,Ca,In} + 0.42(1+2\alpha)(1-x_{W,Ca,In})}{\lambda + 0.21(1+4\alpha)(1-x_{W,Ca,In})} \end{cases} \tag{3.17}$$

The variable α is assumed to be −0.05 for the current at 100 A, 150 A and 200 A. According to the operating conditions, the estimated partial pressure $P_{W,Ca,Out}$ of water vapor is summarized (Table 3.7). Obviously, the estimated partial pressure is lower than the saturated vapor pressure for all current consistently, such that the exhausted air has the capability of carrying more water out of the stack and the stack tends to be dried out. Therefore, the experiment result of the fuel cell system is valid in trend.

The research in this part proves that the proposed method for online AC impedance measurement of the fuel cell system is feasible based on the DC/DC converter but the accuracy of AC impedance needs to be elucidated and improved.

Table 3.7 Estimation of partial pressure of water vapor at cathode outlet (EPPWVCO)

Current (A)	ESR of Air	TACO (°C)	SP (kPa)	EPPWVCO (kPa)	Trend
100	2.3	60	19.932	≤18.79	Drying-out
150	2.28	62	21.968	≤20.63	Drying-out
200	2.2	65	25.022	≤21.24	Drying-out

ESR means the excess stoichiometric ratio. TACO means temperature of air at cathode outlet. SP means the saturated pressure of water vapor

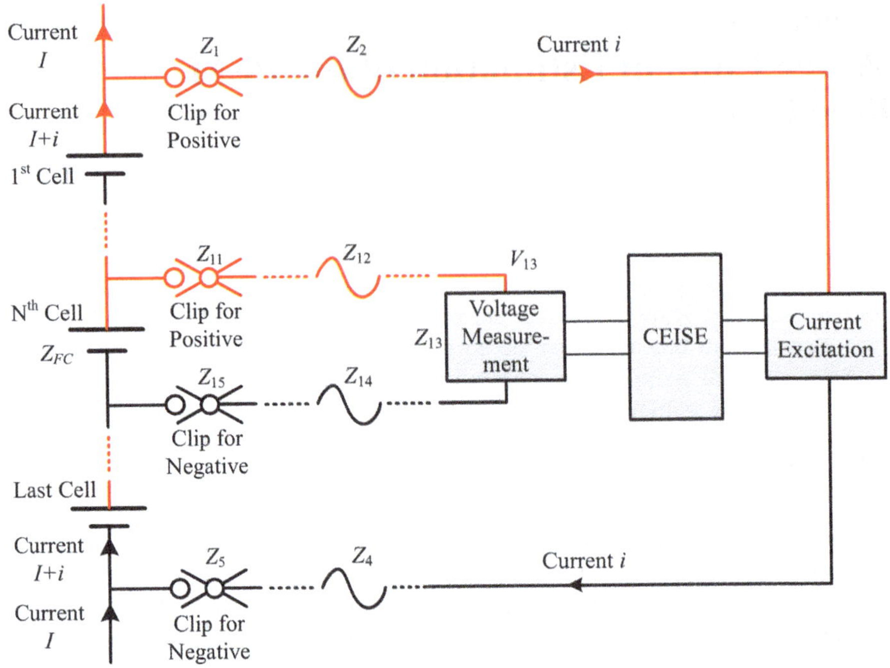

Fig. 4.1 Principle of the electrochemical workstation for AC impedance

Fig. 4.2 Principle of the method for AC impedance by our research

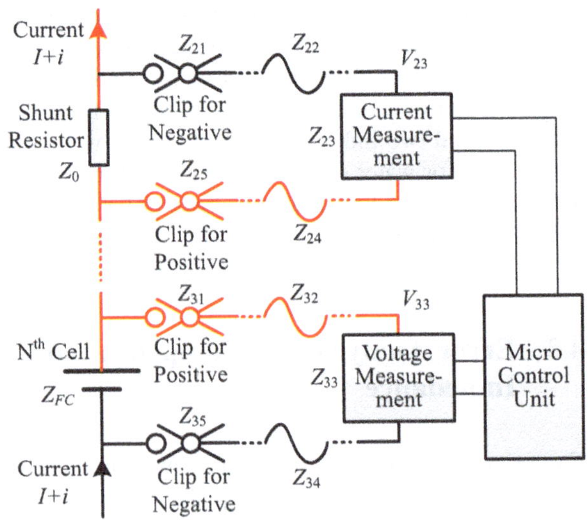

the Nth fuel cell is calculated based on the sampled current excitation and voltage response. In the loop circuit of the current excitation signal, resistance of the positive clip, the positive wire, the negative wire and the negative clip is Z_1, Z_2, Z_3 and Z_4 respectively. In the loop circuit of the voltage response signal of the Nth fuel cell, resistance of the positive clip, the positive wire, the negative wire and the negative clip are Z_{11}, Z_{12}, Z_{14} and Z_{15} respectively. The input resistance of the voltage sampling circuit is Z_{13}. The sum of resistance of the loop circuit of voltage response signal is Z_{10}.

$$V_{13} = i \times \frac{Z_{13}Z_{fc}}{Z_{fc} + Z_{11} + Z_{12} + Z_{13} + Z_{14} + Z_{15}} \tag{4.1}$$

$$\begin{cases} Z_{13} \gg Z_{fc} + Z_{11} + Z_{12} + Z_{14} + Z_{15} \\ V_{13} \approx i \times Z_{fc} \end{cases} \tag{4.2}$$

The measured voltage V_{13} at the input of the voltage sampling circuit is calculated as Eq. 4.1. If the resistance Z_{13} is far large than the rest resistance in the voltage loop circuit, the measured voltage V_{13} is calculated as Eq. 4.2, which means that the influence of the rest resistance on the measured voltage can be neglected.

For our proposal, the specialized DC/DC converter draws current from the stack and the value of the current is sum of stable current I and current excitation i. The drawn current flows through the shunt resistor Z_0. The voltage over the shunt resistor and the voltage response of the Nth fuel cell are measured simultaneously (Fig. 4.2). Part of signal processing circuits for current excitation and voltage response is shown (Fig. 4.3). The resistor R_{spc1}, R_{spc2} and the capacitor C_{spc2} form the low pass filter.

In the first case where the 1st probe and the 2nd probe are connected to a function generator (Fig. 4.3a), for calibration of signal processing circuits (Fig. 3.17), the relation between the voltage V_1, voltage V_2, voltage V_3 and voltage V_4 is presented as Eq. 4.3. The function generator generates a stable voltage and a sinusoidal voltage with small amplitude.

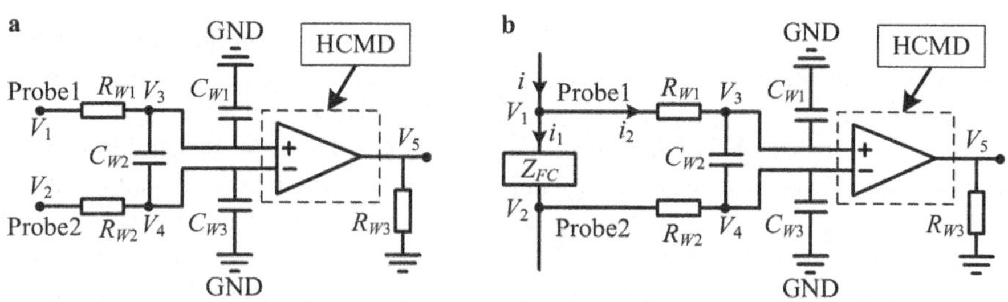

Fig. 4.3 Circuit analysis: **a** Signal processing circuit with a function generator, **b** Signal processing circuit with a fuel cell

$$\begin{cases} \frac{V_1-V_3}{R_{spc1}} = \frac{V_3}{\frac{1}{j\omega C_{spc1}}} + \frac{V_3-V_4}{\frac{1}{j\omega C_{spc2}}} \\ \frac{V_2-V_4}{R_{spc2}} = \frac{V_3-V_4}{\frac{1}{j\omega C_{spc2}}} - \frac{V_4}{\frac{1}{j\omega C_{spc3}}} \end{cases} \tag{4.3}$$

For a resistor and a capacitor, the nominal value and the accuracy are two key parameters. The accuracy of a resistor can be controlled with in 0.1% while that of a capacitor has three levels such as 5%, 10% and 20% and the level at 5% is the most common in application. Even if the capacitor C_{spc1}, C_{spc2} and C_{spc3} are designed to be the same, the relation between the voltage differences (V_3-V_4) and (V_1-V_2) cannot be derived explicitly due to variation of actual parameters. That is to say, the calibration of the signal processing circuit cannot be realized for such configuration in this way.

$$\frac{V_3-V_4}{V_1-V_2} \approx \frac{1}{1+j2\omega R_{spc1}C_{spc2}} \tag{4.4}$$

On the contrary, the relation between the voltage differences (V_3-V_4) and the (V_1-V_2) can be derived approximately as Eq. 4.4 if following conditions are satisfied: the value of capacitor C_{spc2} is far larger than that of the capacitor C_{spc1} and C_{spc3} for at least 3-order magnitude; the value of resistor R_{spc1} is far larger than that of the resistance Z_{11} and Z_{12} for at least 3-order magnitude; the value of resistor R_{spc2} is far larger than that of the resistance Z_{14} and Z_{15} for at least 3-order magnitude. If so, the calibration of the signal processing circuit using the function generator is feasible (Fig. 3.17). Our design satisfies these conditions.

$$\frac{iZ_{fc}}{V_3-V_4} = \frac{\frac{1}{C_{spc1}}+\frac{1}{C_{spc2}}+\frac{1}{C_{spc3}}}{\frac{1}{C_{spc1}}\frac{1}{j\omega C_{spc2}}+\frac{1}{C_{spc2}}\frac{1}{j\omega C_{spc3}}}\left(Z_{fc}+R_{spc1}+R_{spc2}\right)+1 \tag{4.5}$$

$$\frac{iZ_{fc}}{V_3-V_4} \approx \frac{\frac{1}{C_{spc1}}+\frac{1}{C_{spc2}}+\frac{1}{C_{spc3}}}{\frac{1}{C_{spc1}}\frac{1}{j\omega C_{spc2}}+\frac{1}{C_{spc2}}\frac{1}{j\omega C_{spc3}}}\left(R_{spc1}+R_{spc2}\right)+1 = H_V \tag{4.6}$$

In the second case where the 1st probe and the 2nd probe are connected to positive and negative of the Nth fuel cell respectively (Fig. 4.3b). The relation between the voltage differences (V_3-V_4) and (V_1-V_2) can be derived explicitly as Eq. 4.5 and the term on left side of the equation is denoted as H_V. It can be concluded that the design of three capacitors C_{spc1}, C_{spc2} and C_{spc3} has no influence on derivation of H_V and the calibration of the signal processing circuit is feasible. The impedance Z_{fc} of the fuel cell is far smaller than that of the resistor R_{spc1} and R_{spc2} for at least 5-order magnitude, such that H_V is approximately calculated as Eq. 4.6. The influence of resistance of wire and clip exists on H_V can also be neglected due to big difference in magnitude of their value. Interestingly, when value of capacitor C_{spc2} is far larger than that of the capacitor C_{spc1} and C_{spc3} for at least 3-order magnitude, Eq. 4.4 is equal to Eq. 4.6.

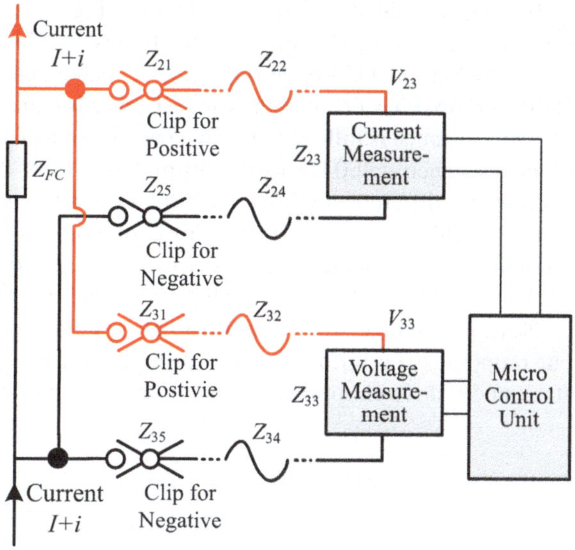

Fig. 4.4 Method for calibration of the cell voltage monitor

The transfer function of the signal processing circuit from the input of the high common-mode differential circuit to the analog input of the microprocessor, as Fig. 3.15 and Eq. 3.16, is the same for both the current excitation signal and the voltage response of a fuel cell, but their amplitude attenuation and phase delay are different. They are denoted as H_{V1} and H_{CV1} respectively. For accurate calibration of signal processing circuits in the cell voltage monitor, the experiment is proposed (Fig. 4.4).

$$\begin{cases} \frac{iZ_{fc}}{(V_3-V_4)_{CV}} = H_{CV}\frac{Z_{fc}+Z_{20}}{Z_{20}} \approx H_{CV} \\ V_{CAD} = (V_3 - V_4)_{CV} H_{CV1} \end{cases} \quad (4.7)$$

$$\begin{cases} \frac{iZ_{fc}}{(V_3-V_4)_{V}} = H_{V}\frac{Z_{fc}+Z_{30}}{Z_{30}} \approx H_{V} \\ V_{VAD} = (V_3 - V_4)_{V} H_{V1} \end{cases} \quad (4.8)$$

The resistance Z_{20} is sum of the resistance Z_{21}, Z_{22}, Z_{23}, Z_{24} and Z_{25} and the voltage at the analog input of the microprocessor is V_{CAD}. Similarly, the resistance Z_{30} is sum of the resistance Z_{31}, Z_{32}, Z_{33}, Z_{34} and Z_{35} and the voltage at the analog input of the microprocessor is V_{VAD}. The transfer function of the signal processing circuit for the current excitation signal is presented as Eq. 4.7 and that for the voltage response of a fuel cell presented as Eq. 4.8. If the value of resistance Z_{20} and that of resistance Z_{30} are far larger than that of resistance Z_{fc}, the interaction between Z_{20} and Z_{30} can be neglected as Eqs. 4.7 and 4.8. The experiment result of the difference between the transfer function H_{CV1}/H_{CV} and H_{V1}/H_V is 1.052 for the amplitude attenuation and −2.79° for phase delay with current excitation at 400 Hz.

Besides, the shunt resistor is calibrated (Fig. 4.5) and its resistance varies with frequency (Fig. 4.6). The room temperature is 25 °C in the constant temperature test station by SHENLI Corporation. The electrochemical workstation is the Reference 3000 by GAMRY Corporation. The resistance is equal to a resistor at low frequency but the resistance is influenced by the parasite inductance. The amplitude and phase of the resistance at 400 Hz are 0.2691 mΩ and 20.88° respectively. Difference of the resistance at different amplitude of current excitation is within 0.3% at all frequency and it's highly stable. Except for this calibration, the influence of temperature on resistance of the shunt resistor must also be considered and the resistance is increased by 9.26% at 65 °C in a controllable temperature chamber.

Next, the comparison experiment between the electrochemical workstation and our proposal is performed by measuring the impedance of the same fuel cell in a stack under different operating conditions. The fuel cell stack has 10 cells and the

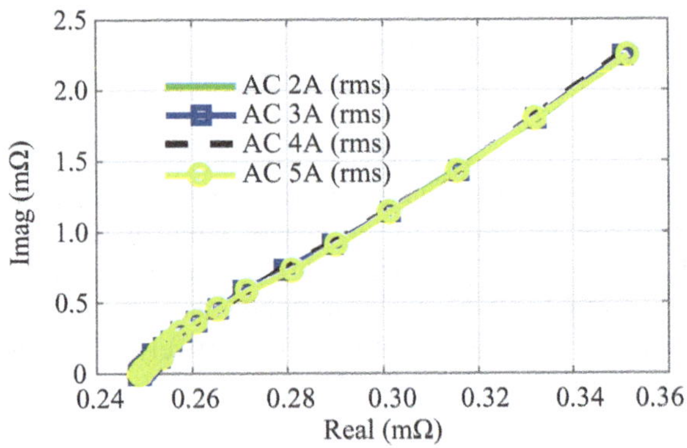

Fig. 4.5 Method for calibration of the shunt resistor

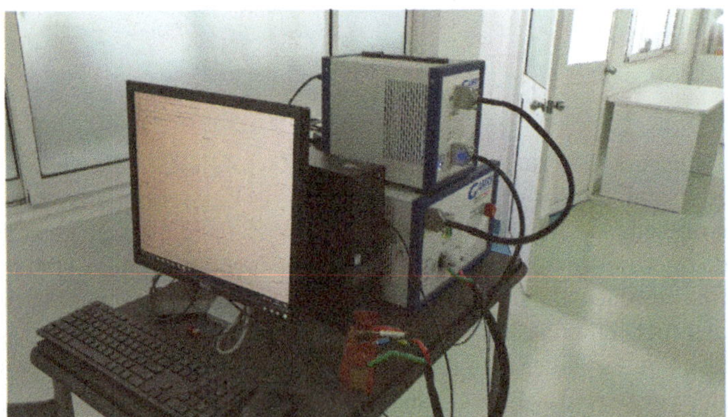

Fig. 4.6 Resistance of the shunt resistor with frequency

Table 4.1 Measured resistance of the fuel cell

Freq. (Hz)	RH (%)	Res. Amp. (mΩ)	Res. Phase (°)	Amp. Diff. (%)	Phase Diff. (°)
400	Ca 80/An 80	0.3438 (ECWS)	−12.94	/	/
400	Ca 80/An 80	0.3336 (AOR)	−12.04	−2.98	0.9
1126	Ca 80/An 80	0.3112 (ECWS)	−0.06	/	/
400	Ca 20/An 20	0.3573 (ECWS)	−12.80	/	/
400	Ca 20/An 20	0.3666 (AOR)	−11.32	2.60	1.48
1126	Ca 20/An 20	0.3211 (ECWS)	0.34	/	/
400	Ca 80/An 80	0.3471 (ECWS)	−12.88	/	/
400	Ca 80/An 80	0.3367 (AOR)	−12.11	−3.01	0.77
1126	Ca 80/An 80	0.3137 (ECWS)	−0.08	/	/
400	Ca 20/An 20	0.3588 (ECWS)	−12.96	/	/
400	Ca 20/An 20	0.3709 (AOR)	−11.54	3.38	1.42
1126	Ca 20/An 20	0.3234 (ECWS)	−0.81	/	/

RH means the relative humidity of supplied air at cathode inlet and supplied hydrogen at anode inlet. ECWS means the electrochemical workstation. AOR means the approach by our research

effective operating area is 270 cm^2 by SHENLI Corporation, Reference 3000 is still our choice and the test plant is G400 by GREENLIGHT Corporation from Canada. A new DC/DC converter is designed to be suitable for such low operating voltage and the CVM is the same as in Chap. 3.

The output current of the stack is 150 A, the coolant temperature at the coolant inlet of the stack is 65 °C and the flow rate is constant. The relative humidity of the supplied air and hydrogen varies from 80 to 20% and then from 20 to 80% again but variation of their relative humidity is not simultaneous. For each relative humidity, the operating condition is maintained for 30 min. Each time once impedance of the fuel cell is measured by our proposal at 400 Hz, the electrochemical workstation is activated to measure the impedance at frequency from 10 kHz to 1 Hz. The experiment result is listed (Table 4.1).

The amplitude difference is defined as the difference, which is equal to amplitude of the impedance by our proposal minus that of the impedance by the workstation, divided by that of the impedance by the workstation. The phase difference is equal to the phase of the impedance by our proposal minus that of the impedance by the workstation.

It shows that the amplitude difference is about 3% larger for low relative humidity of gas supply and it's about 3% smaller for high relative humidity of gas supply. The phase difference is within 2° definitely.

Besides, the impedance spectroscopy of the fuel cell by the electrochemical workstation is illustrated (Fig. 4.7). The impedance at high frequency range, on the left-half side, is stable while that at low frequency range is dispersed so it's no appropriate to measure the low frequency impedance for further analysis. The enlarged Nyquist plot for the high frequency range shows that the increased relative humidity results

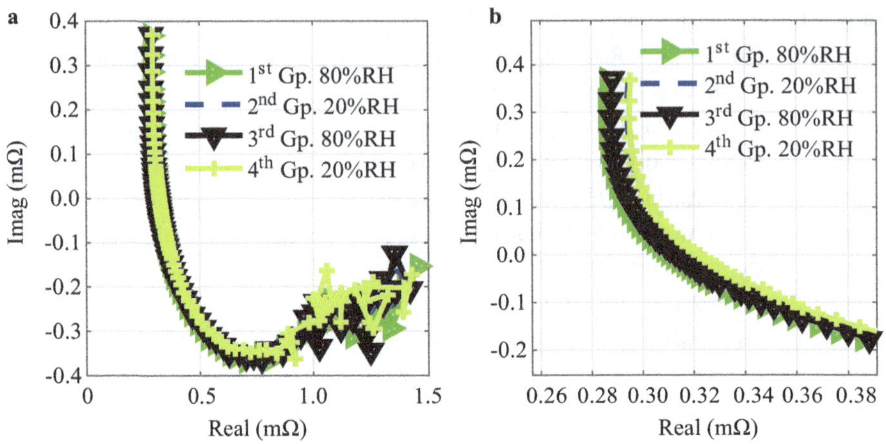

Fig. 4.7 **a** Effect of relative humidity on EIS of the fuel cell, **b** Locally enlarged EIS of the fuel cell

in decrease of the high frequency impedance whose phase is zero. At most frequency lower than 1 kHz, the impedance is in the fourth phase of Nyquist plot and the fuel cell shows capacitor characteristic. When the frequency is larger than 1,126 Hz, the impedance is in the first phase of Nyquist plot and the fuel cell shows inductor characteristic.

According to data in Table 4.1, the phase of resistance at 400 Hz is not zero and it's not the high frequency resistance exactly. However, the slope of the connected line between the resistance at 400 Hz and the resistance at 1,126 Hz keeps around 3.1 and the variation is within $\pm$ 4%. When the relative humidity of supplied air changes from 80 to 20%, the high frequency resistance is increased only by 0.01 mΩ, which proves the capability of maintaining water content of the fuel cell with porosity of the gas diffusion layer only 0.288. The experiment for current density at 100, 500, 800 and 1300 mA/cm^2 shows the similar trend.

In a word, the proposed method for error analysis and compensation for the online AC impedance is valid and the measured AC impedance by our proposal is reliable and of high accuracy. The water content estimation will be based on the measured AC impedance by our approach.

4.3 Sensitivity Analysis of Fuel Cell with Self-humidifying MEA

The self-humidifying MEA is realized by enhancing the hydrophilic property and lowering the effective porosity of the gas diffusion layer, such that more water can be stored in the MEA to prevent the fuel cell from being dried out. In terms of the fuel cell system, the external humidifier may be eliminated. The disadvantage is that the oxygen transport resistance is greatly increased and to maintain high performance, the operating temperature and gas pressure shall be increased. During purging processing when the fuel cell system needs to stop, the stored water can be removed but with longer purging time. Some fuel cell systems have adopted this kind of MEA and the water content estimation for them is necessary.

4.3.1 Analysis Based on High Frequency Resistance

The fuel cell stack with 10 cells in serial is our research object by SHENLI Corporation and the test plant is G400 by GREENLIGHT Corporation. It is equipped with self-humidifying MEA. The air stoichiometric ratio is 2.5, the hydrogen stoichiometric ratio 1.5 and the operating temperature 65 °C. For each current density, each time after changing the relative humidity of supplied air or supplied hydrogen, the operating condition is maintained for 30 min and then the high frequency resistance is measured by Reference 3000 by GAMRY Corporation (Table 4.2).

When the relative humidity of supplied air is increased from 0 to 30%, the HFR at 0.1 A/cm^2 is decreased by 0.04 mΩ while that at higher current density is decreased by only 0.01 mΩ. When the relative humidity of supplied air is increased from 30 to 100%, the HFR at all current density is decreased by only 0.01 mΩ. When the current density is no less than 0.5 A/cm^2, the HFR is decreased by about 6% for most cases. The voltage of the fuel cell, resulting from the decreased HFR by increased relative humidity of supplied air, is only increased by 6.3 mV with effective surface area at

Table 4.2 HFR at 400 Hz of the fuel cell and operating condition

RH (%)	CD (A/cm^2)	HFR Freq. (Hz)	Res. Real (mΩ)	Res. Imagine (mΩ)
An 80/Ca 0	0.1	1266	0.3479	−0.004
An 80/Ca 0	0.5	998	0.3125	0.0019
An 80/Ca 0	0.8	998	0.3219	0.0015
An 80/Ca 0	1.3	998	0.3301	−0.0025
An 80/Ca 30	0.1	998	0.3063	0.004
An 80/Ca 30	0.5	998	0.2986	0.0001
An 80/Ca 30	0.8	998	0.3098	0.0014
An 80/Ca 30	1.3	998	0.3238	0.0051
An 100/Ca 100	0.1	998	0.2945	0.0007
An 100/Ca 100	0.5	895	0.2899	0.0001
An 100/Ca 100	0.8	998	0.2973	−0.0053
An 100/Ca 100	1.3	998	0.3121	0.0012

RH means the relative humidity of supplied air at cathode inlet and supplied hydrogen at anode inlet. CD means the current density

270 cm^2. When the current density is lower, the contribution of increasing relative humidity of supplied air to the voltage of a fuel cell is decreased. It's concluded that the self-humidifying MEA is not sensitive to the relative humidity of supplied air but external humidifier can still increase water content in the MEA to some extent.

Then the approach to online AC impedance measurement proposed in Chap. 3 is utilized to measure the HFR at 400 Hz for fuel cells and the operating conditions are varied in procedure (Fig. 4.8). In the first group experiment, the current density is 0.62 A/cm^2, the air stoichiometric ratio 2.5, the hydrogen stoichiometric ratio 1.5 and the operating temperature 65 °C. To be noticed is that the pipe between the test plant and the fuel cell stack is relatively long and actual operating conditions of the fuel cell stack are slightly different from the data (Fig. 4.8). For 7,500 to 8,300 s, the fuel cell operates in steady state but the output pressure at cathode outlet varies

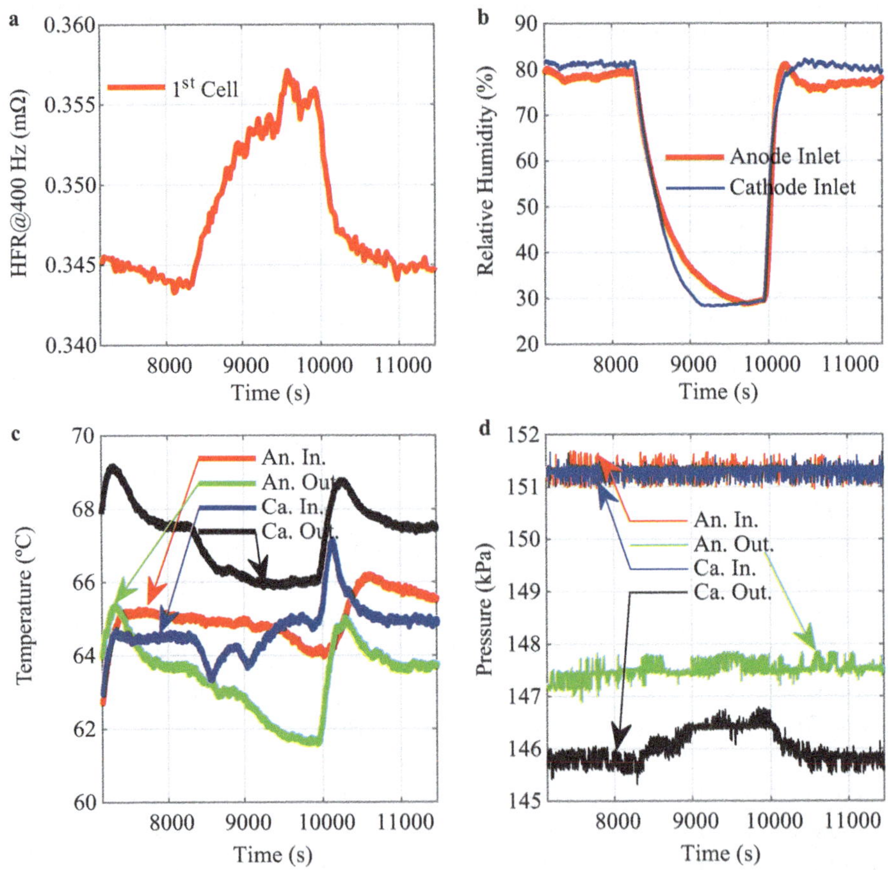

Fig. 4.8 The first group: **a** Resistance at 400 Hz of the 1st cell, **b** Relative humidity of supplied gas at anode and cathode inlet, **c** temperature of gas at inlet and outlet of anode and cathode, **d** pressure of gas at inlet and outlet of anode and cathode

between 145.5 and 146 kPa. From 8,300 to 9,500 s, the relative humidity of supplied air and hydrogen is decreased from 80 to 30% but the HFR at 400 Hz of the first cell is just increased since 8,335 s. The output pressure at cathode outlet is increased and then it keeps stable since 9,000 s, because the relative humidity of supplied air reaches around 30%.

The relative humidity of supplied air is decreased at faster speed than that of the relative humidity of supplied hydrogen. From 9,000 s to 9,500 s, even though the relative humidity of supplied hydrogen is decreased, the HFR is oscillated around 0.353 mΩ with no obvious increase or decrease. From 9,300 s to 9,700 s, the output pressure at the anode outlet is increased for some time and then it recovers. Since the relative humidity of both supplied air and hydrogen reaches 30% at 9,500 s, the HFR starts to vibrate around 0.355 mΩ and it ranges from 0.353 mΩ to 0.357 mΩ. It may result from decreased output temperature at the anode outlet and increased input temperature at the cathode inlet.

The relative humidity of both supplied air and hydrogen is increased from 30 to 80% and the HFR at 400 Hz is decreased at a faster speed than that of the opposite situation. After their relative humidity reaches 80%, the output pressure at anode outlet and cathode outlet oscillate intermittently. It may result from the slug and removal of liquid water in the flow channel. At the same time, the HFR is decreased gradually from 0.346 mΩ to 0.344 mΩ.

Based on the one-dimensional steady-state fuel cell model as Sect. 2.3, the proton-conduction resistance in the membrane is calculated and the cathode and anode catalyst layers are considered to be part of the membrane. The equivalent thickness of the membrane is assumed to range from 15 μm to 45 μm to show their difference. The actual thickness of the cathode catalyst layer, the anode catalyst layer and the membrane is 15 μm. The simulation result according to the operating conditions is illustrated (Fig. 4.9). The HFR is increased much larger for each membrane thickness than that of the experiment result. There are maybe two reasons: parameters of the

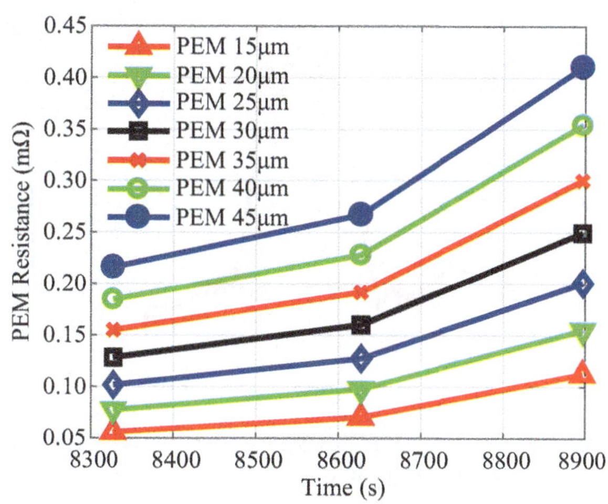

Fig. 4.9 PEM resistance with time with different thickness

membrane and the catalyst layer are all from old references and no updated data is available but for the last two decades the material for the fuel cell has been advanced in a surprising speed; the capability of water storage of a fuel cell with self-humidifying MEA is strong enough to be less sensitive to operating conditions when it keeps running.

The conclusion of the first group experiment is that water content estimation based on the HFR and the one-dimensional steady-state fuel cell model is not accurate due to unknown material parameters and other uncertainty. In the following parts, the research will be focused on the sensitivity of the fuel cell to operating conditions, including the relative humidity and stoichiometric ratio of supplied air.

The second group experiment is performed with constant relative humidity of suppled hydrogen and the stoichiometric ratio is 2.28 through this experiment. Other procedure is the same as that of the first group. The experiment result is shown (Fig. 4.10). From 27,800 to 28,300 s, the relative humidity of supplied air is decreased from 80 to 30% and the HFR is increased with approximate constant slope with 40 s time delay.

In comparison with the result of the first group, the HFR still oscillates between 0.354 mΩ to 0.358 mΩ when the relative humidity of supplied air is stable but the average HFR won't be further increased. The HFR is increased by 0.01 mΩ for the second group but the HFR is increased by 0.012 mΩ. The difference results from lowered capability of water drainage with decreased air supply flow rate.

The third group experiment is performed with alternatively changed relative humidity of supplied air and hydrogen (Fig. 4.11). Firstly, the relative humidity of supplied hydrogen is decreased from 37,300 s to 38,500 s and the HFR is increased by 0.002 mΩ with time delay. After the relative humidity of supplied hydrogen is stable, the HFR keeps constant from 38,500 s to 38,900 s.

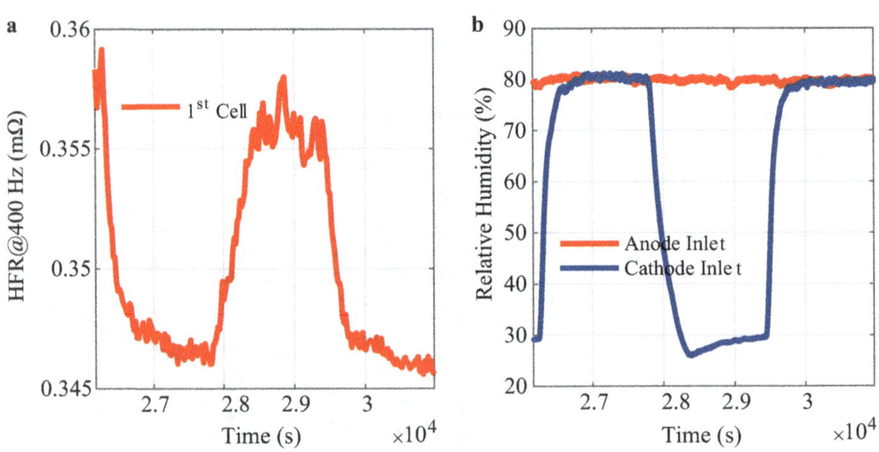

Fig. 4.10 The second group: **a** Resistance at 400 Hz of the 1st cell, **b** Relative humidity of supplied gas at anode and cathode inlet

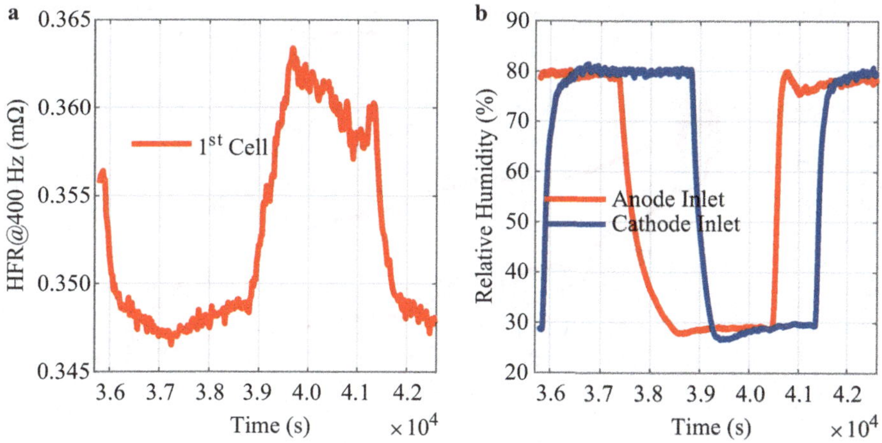

Fig. 4.11 The third group: **a** Resistance at 400 Hz of the 1st cell, **b** Relative humidity of supplied gas at anode and cathode inlet

Secondly, the relative humidity of supplied air is decreased from 38,900 s to 39,700 s and the HFR is increased by 0.014 mΩ with time delay. The HFR reaches the maximal at 39,700 s and then it decreases gradually until 40,300 s. Thirdly, from 40,300 to 41,200 s, the relative humidity of supplied hydrogen is increased from 30 to 80% and the HFR experiences the process of being decreased, being stable and being increased and being stable in sequence and finally it becomes stable when the relative humidity is stable. Fourthly, the relative humidity of supplied air is increased from 30 to 80% since 41,300 s, the HFR is decreased at almost constant speed at first and then it decreases gradually.

This experiment demonstrates that the air supply has greater influence on the HFR than the hydrogen supply and humidification of supplied hydrogen is important but a high relative humidity has no large contribution to water content. The most effective method of removing water content in a fuel cell is to adjust the relative humidity of supplied air. When the current density is 0.44 A/cm^2 and the operating temperature is 60 or 65 °C, the experiment result is similar with the same test procedure.

When the current density is 0.62 A/cm^2 and the operating temperature 65 °C, the relation between relative humidity of supplied air and the HFR at 400 Hz which is measured every 30 s is plotted for two stoichiometric ratios respectively (Fig. 4.12).

The result shows that water removing is slow by decreasing relative humidity of supplied air but the water restoring is fast by increasing its relative humidity, based on the concentrated points of measured HFR. The reasons are as follows: removing water from MEA is in a passive way of increasing gradient of water content concentration and this gradient enhances water transport from MEA to gas diffusion layer with time delay; restoring water inside the MEA is in an active way of decreasing the gradient of water concentration and the generated water in the MEA cannot be transported directly, so water accumulates in the MEA without any time delay. When the air

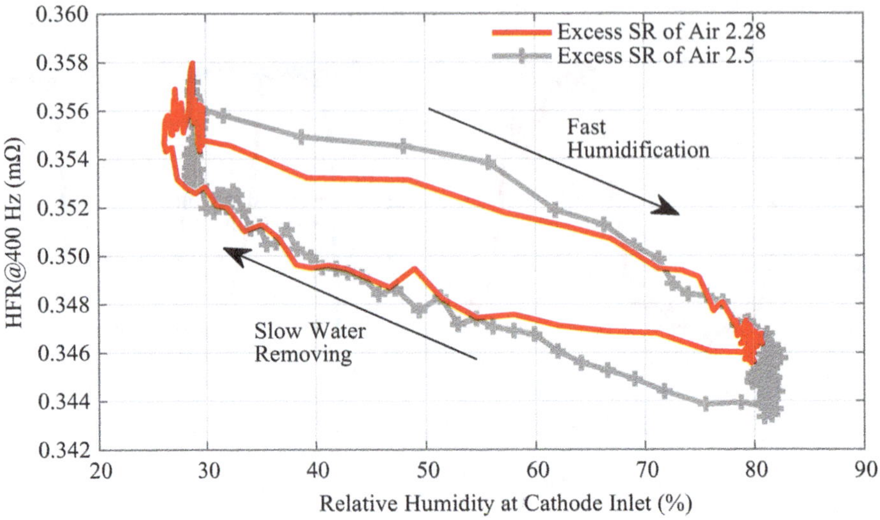

Fig. 4.12 Dynamic resistance at 400 Hz of the 1st cell under different excess ratio of supplied air

stoichiometric ratio is decreased, the capability of removing water from MEA is decreased while that of restoring water inside MEA is increased. The closed loop of the variation of the HFR at 400 Hz is smaller for the operating conditions with lower air stoichiometric ratio.

Based on the three group experiment and part of the experiment result in Table 4.2, the HFR at 400 Hz for each current density is plotted together (Fig. 4.13). The air stoichiometric ratio and the operating temperature are the same for these current densities. When the current density is 0.1 A/cm^2, the increase from 0 to 30% of relative humidity of supplied air leads to significant decrease of the HFR and in other current density or relative humidity the HFR is decreased at similarly slope. For all current density, the increase of relative humidity of supplied air leads to decrease of the HFR. When the relative humidity of supplied air is the same, the

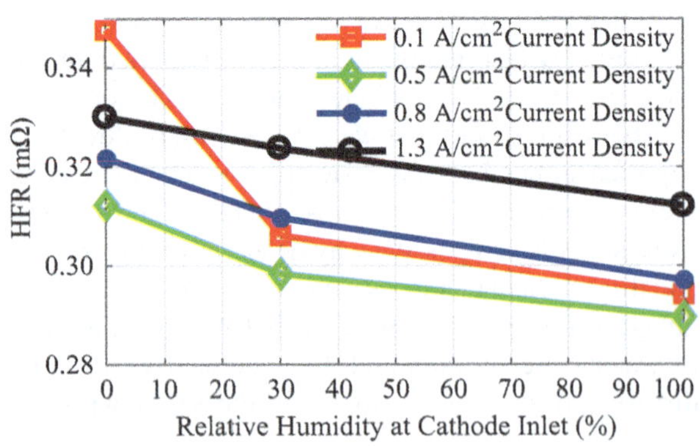

Fig. 4.13 Resistance at 400 Hz of the 1st cell under different relative humidity of supplied air and at different current density

HFR at 0.5 A/cm^2 current density is the lowest and the HFR is increased when the current density is higher than 0.5 A/cm^2.

Theoretically, when the relative humidity of supplied air and hydrogen is 100%, water content in the MEA ought to be saturated and the HFR should be the at the maximal. However, the varied HFR with current density demonstrates that the MEA is not full of liquid all the time and the content of liquid water in the MEA changes under different operating conditions. It further proves that the modeling of the MEA in Sect. 2.4 is reasonable.

For the fuel cell with self-humidifying MEA, the HFR is the most effective way to analyze the sensitivity of the fuel cell performance to operating condition. Even though the exact relation between the HFR and the water content in MEA is not known, it can semi-quantitatively indicate the water content.

4.3.2 *Water Content with Respect to the Fuel Cell Voltage*

The overall content of the fuel cell is sum of water content in MEA, water content in the gas flow channel, and water content in the gas diffusion layer. Though the high frequency resistance reflects water content in the MEA directly, measuring the high frequency resistance in the fuel cell system all the time is not appropriate due to power excitation to the powertrain system and the possible energy loss in the DC/DC converter. Besides, the coordination of the fuel cell system controller, the cell voltage monitor and the DC/DC converter is complex and other methods can be adopted to relieve such high control difficulty.

$$\begin{cases} \sigma_{avg} = \frac{1}{N_{stack}} \sum_{m=1}^{N_{stack}} V_m \\ V_{StdD} = \sqrt{\frac{1}{N_{stack}-1} \sum_{m=1}^{N_{stack}} \left(V_m - \sigma_{avg}\right)^2} \end{cases} \tag{4.9}$$

$$\Delta V_{\max} = \max_{m,n}\{V_m - V_n\} \tag{4.10}$$

According to result in Sect. 4.3.1, the water content of a fuel cell changes slowly for hundreds of seconds and the output voltage of the fuel cell changes at the same time. The output voltage is function of overall water content of the fuel cell. The index based on statistical analysis of the output voltage is presented as Eq. 4.9, whereas σ_{avg} is the average cell voltage of the fuel cell stack and V_{StdD} the standard deviation. The voltage difference ΔV_{max} between the highest cell voltage and the lowest cell voltage is another index as Eq. 4.10, whereas m and n are the number of two difference cells and V_m is voltage of the mth fuel cell.

Experiment is performed on a fuel cell stack with 120 cells in serial and the test plant is G600 by GREENLIGHT Corporation (Fig. 4.14). The current density is 0.75 A/cm^2, the operating temperature 65 °C, the effective surface area 270 cm^2 and the

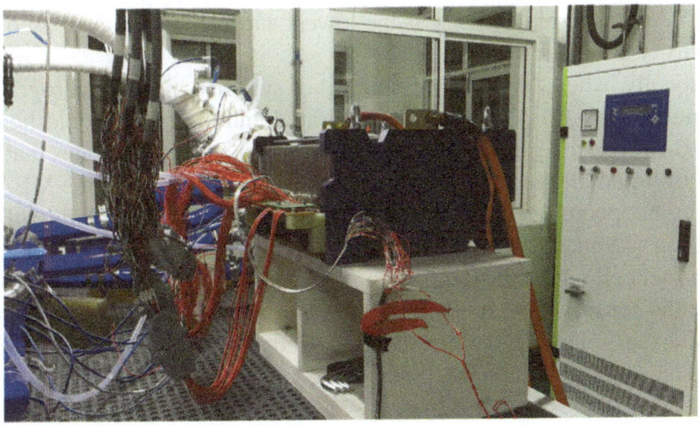

Fig. 4.14 Test plant for the fuel cell stack

stoichiometric of supplied air and supplied hydrogen is 2.5 and 1.5 respectively. The DC/DC converter and the cell voltage monitor together measures the high frequency resistance and the cell voltage monitor measures the cell voltage as well for every 300 ms.

The test is performed as following procedure: firstly, the relative humidity of supplied air and supplied hydrogen is changed simultaneously for four times; secondly, the relative humidity of supplied air and that of supplied hydrogen are changed alternatively for four times. The experiment result is illustrated (Fig. 4.15) and the voltage standard deviation and the voltage difference shows different trend.

Firstly, when the relative humidity of supplied air and supplied hydrogen is decreased, the voltage standard deviation decreases synchronously but the voltage difference is increased significantly. When the relative humidity of both gases becomes stable, the voltage standard deviation also maintains around a constant with a little oscillation. When the relative humidity of supplied air is decreased with that of supplied hydrogen constant, the voltage standard deviation is decreased with almost the same slope versus time and then it is in a stable state. However, the trend of the voltage difference is not very convincing with small variation due to high relative humidity of supplied hydrogen.

Secondly, when the relative humidity of supplied air and supplied hydrogen is increased, the voltage standard deviation increases synchronously but with smaller slope versus time. After the relative humidity of both supplied gases is stable, the index keeps growing. On the contrary, the voltage difference is decreased quickly for some time and then it oscillates apparently around 102 mV. When the relative humidity of supplied air is increased with that of supplied hydrogen constant, the voltage standard deviation is increased with larger slope versus time and then it increases gradually. Again the trend of the voltage difference is not persuasive and its oscillation is obvious.

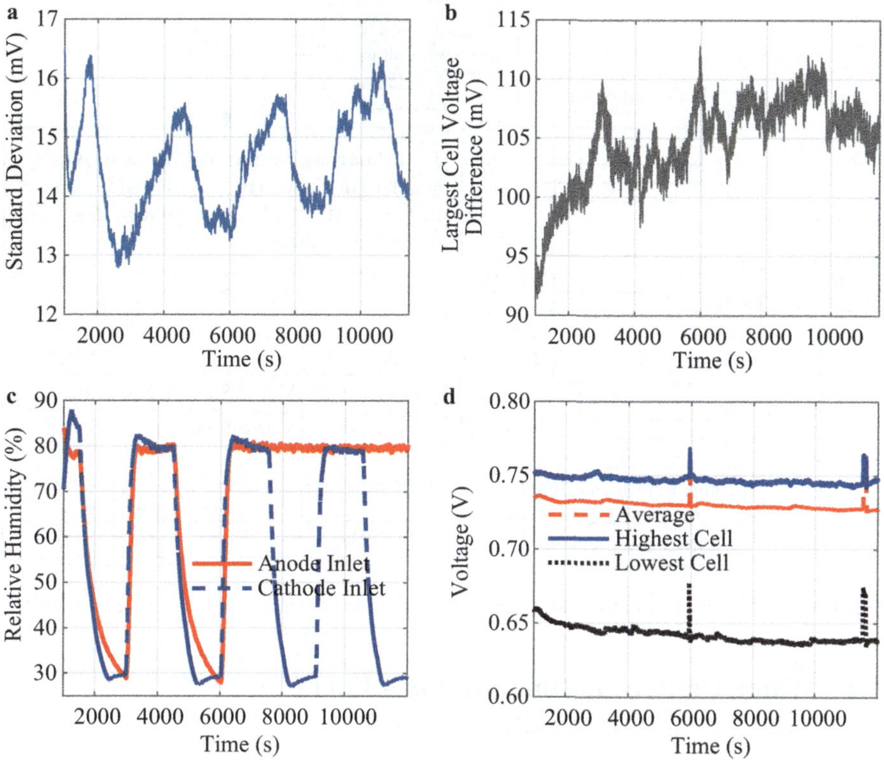

Fig. 4.15 **a** Standard deviation of cell voltage, **b** The largest cell voltage difference, **c** Relative humidity of supplied gas at anode and cathode inlet, **d** The maximal, minimal and average cell voltage

Thirdly, when the relative humidity of supplied air and supplied hydrogen keeps high for a long time, the highest cell voltage is stable but the lowest cell voltage oscillates and decreases slowly. When the relative humidity of supplied air and supplied hydrogen keeps low for a long time, the highest cell voltage, the lowest cell voltage and the average cell voltage are all stable.

That is to say, when the fuel cell is flooding, the uniformity of cells in the stack is worsened and the lowest cell voltage is not stable due to over-accumulated water content inside. When the fuel cell is drying out, the uniformity of cells in the stack is improved and the lowest cell voltage is decreased more severely. The statistical indexes of the cell voltage in the fuel cell contributes to qualitative judgement for trend of water content in the stack and they can act as the trigger for measuring the high frequency resistance (Fig. 4.16). In this way, the system control is partially simplified.

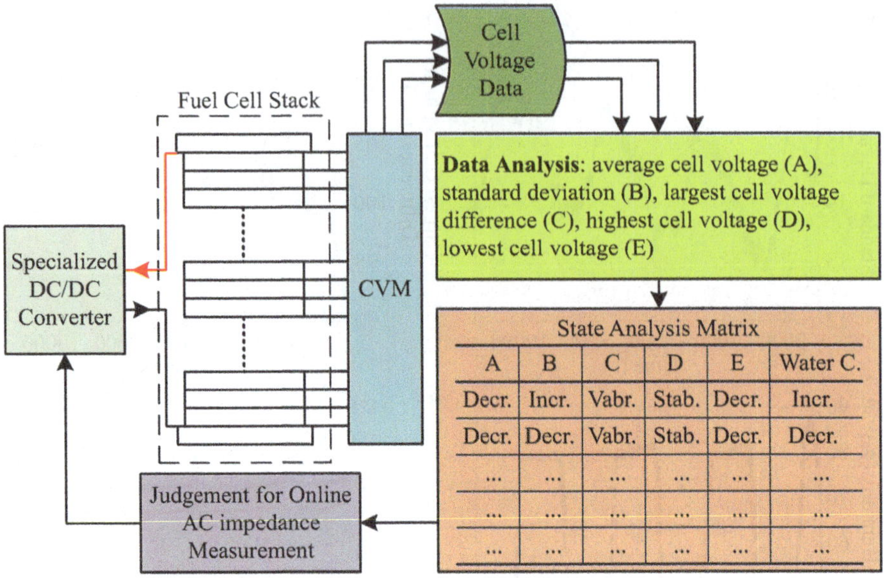

Fig. 4.16 Method for qualitative judgement for trend of water content and the trigger for HFR

4.4 Water Content Estimation for a Fuel Cell with External-Humidifying MEA

4.4.1 Average Water Content of MEA

The water content of a fuel cell with external-humidifying MEA is significantly influenced by the operating conditions like temperature and relative humidity of supplied gases. Besides, the supplied hydrogen is usually not humidified but water is transported from the cathode to the anode. Based on the simplified MEA model with catalyst layer treated as an extension of membrane, the high frequency resistance is on one hand used to calculate the average water content (Fig. 4.17). On the other hand, the actual water content is calculated based on local water content parameter at the interface of the cathode gas diffusion layer and the cathode catalyst layer and that at the interface of the anode gas diffusion layer and the anode catalyst layer.

The contact resistance at interfaces and the electronic-conduction resistance in the bipolar plates and the gas diffusion layer keep stable under the same operating condition. The high frequency resistance minus the contact resistance and the electronic-conduction resistance is the proton-conduction resistance. Comparison is made between the average and the actual water content.

The operating temperature is 70 °C and the thickness of MEA 45 μm. The simulation result is illustrated (Fig. 4.18). When the current density is increased, the

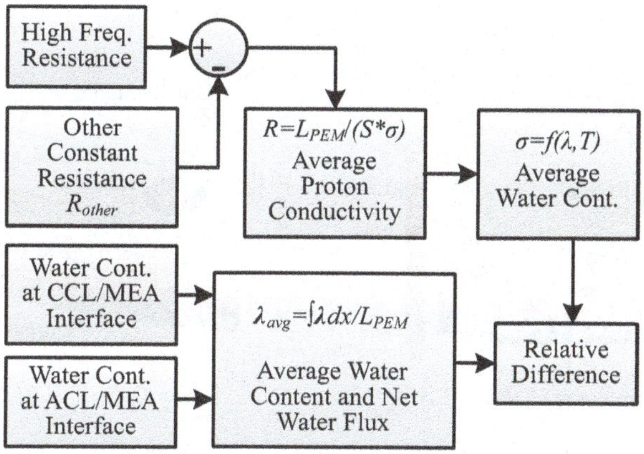

Fig. 4.17 Average water content in MEA based on HFR and that based on PEM model

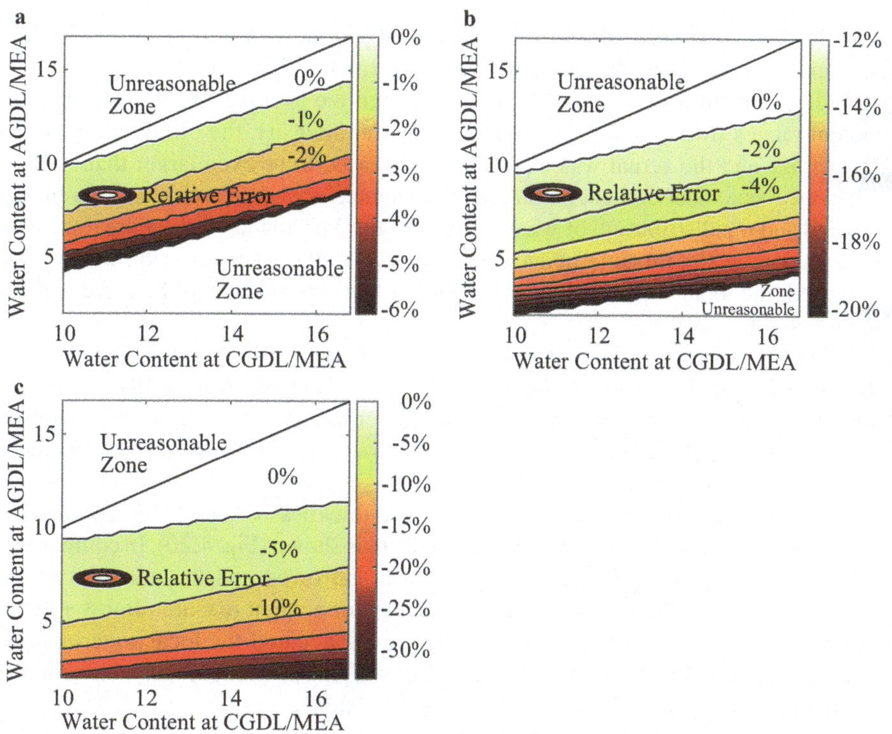

Fig. 4.18 **a** Difference between average and actual water content at 0.4 A/cm^2, **b** Difference between average and actual water content at 0.8 A/cm^2, **c** Difference between average and actual water content at 1.2 A/cm^2

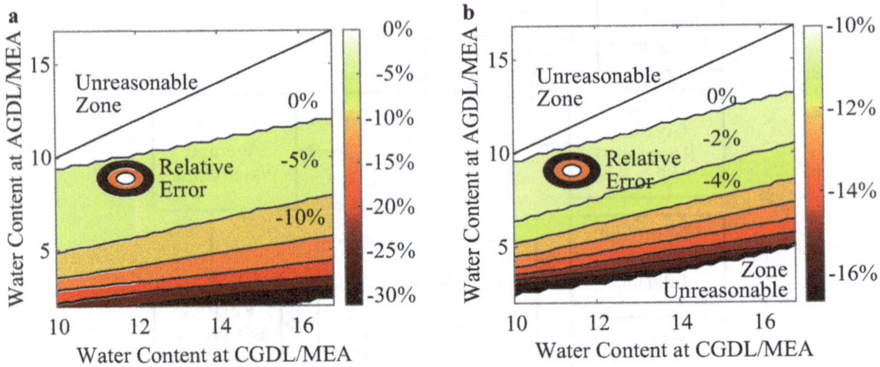

Fig. 4.19 **a** Difference between average and actual water content at 60 °C, **b** Difference between average and actual water content at 80 °C

unreasonable zone is narrowed. The unreasonable zone is where the net water flux coefficient through the MEA is larger than 1 or smaller than -1.

At low current density, the difference between the average and the actual water content ranges from -7% to 0%. At higher current density, the difference between the average and the actual water content is increased to be even larger than -10%. When the local water content parameter at both interfaces increases, the difference is reduced and the distribution of water content in the MEA tends to be more even with decreased net water flux coefficient. To be noticed is that when the current density is increased, the local water content parameter at both interfaces will be increased for an operated fuel cell.

When the current density is 0.8 A/cm^2, the thickness of MEA 45 μm and the operating temperature 60 or 80 °C, the simulation result is shown (Fig. 4.19). Compared with that (Fig. 4.18), the unreasonable zone is enlarged with a higher temperature but the maximal of the difference between the average and the actual water content is decreased significantly.

When the current density is 0.8 A/cm^2, the thickness of MEA 55 μm and the operating temperature 70 °C, the simulation result is shown (Fig. 4.20). In comparison with that (Fig. 4.18), the unreasonable zone is narrowed and the maximal of the difference between the average and the actual water content reaches almost -30%.

The relation between the high frequency resistance and the local water content parameters at both interfaces is presented (Fig. 4.21). The higher local water content at any interface, the lower the high frequency resistance.

For all cases, the higher the local water content parameter at the interface of the anode catalyst layer and the anode gas diffusion layer, the smaller difference between the average and the actual water content. When the operating temperature is increased or the thickness of MEA is decreased, the calculated average water content based on only the high frequency resistance is more accurate. To improve the accuracy, the method for estimating the net water flux coefficient is required.

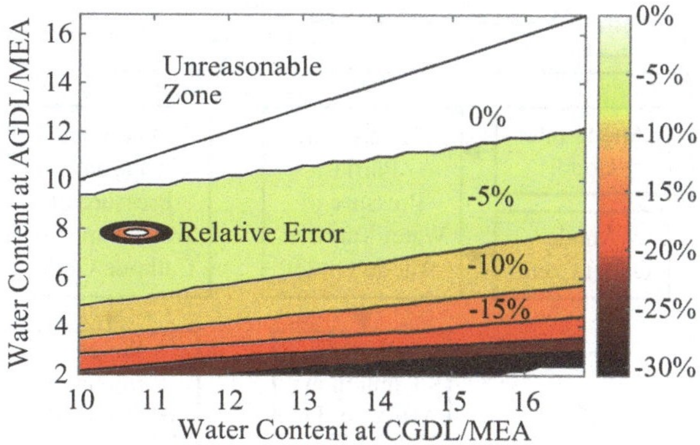

Fig. 4.20 Effect of membrane thickness on difference between average and actual water content

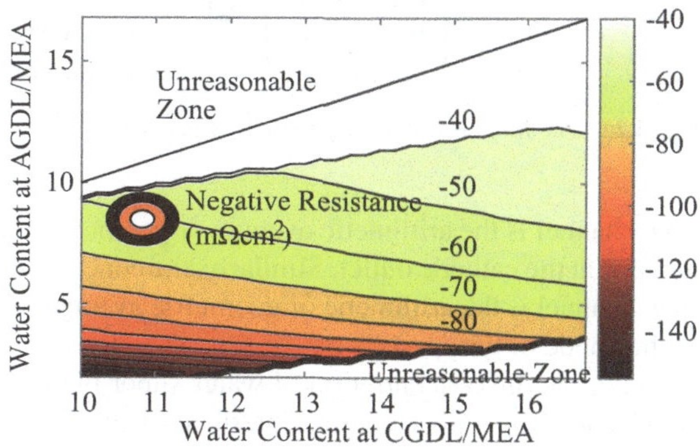

Fig. 4.21 PEM resistance with different water content at AGDL/MEA and at CGDL/MEA

4.4.2 Distribution of Water Content Through the MEA

The distribution of water content through the MEA depends on the net water flux coefficient. It's assumed that the distribution of water content is linear through the MEA in Ref. [1] and the governing equation for water transport from the cathode to the anode is presented as Eq. 4.11 based on the electro-osmotic drag and the back diffusion of water. The electro-osmotic drag coefficient n_{drag} and the effective diffusion coefficient $D_{W,MEA,eff}$ is function of water content in MEA. The back diffusion originates from the water concentration in the cathode flow channel and that in the anode flow channel.

$$\beta \frac{I}{F} = M_W N_{fc} \left(\frac{n_{drag} I}{F} - D_{W,MEA,eff} \frac{c_{W,Ca} - c_{W,An}}{t_{mem}} \right) \tag{4.11}$$

Firstly, the estimation of net water flux coefficient for a fuel cell in flooding state is presented (Fig. 4.22). In the one-dimensional steady state model of the water transport through the membrane, the concentration of water vapor in the cathode

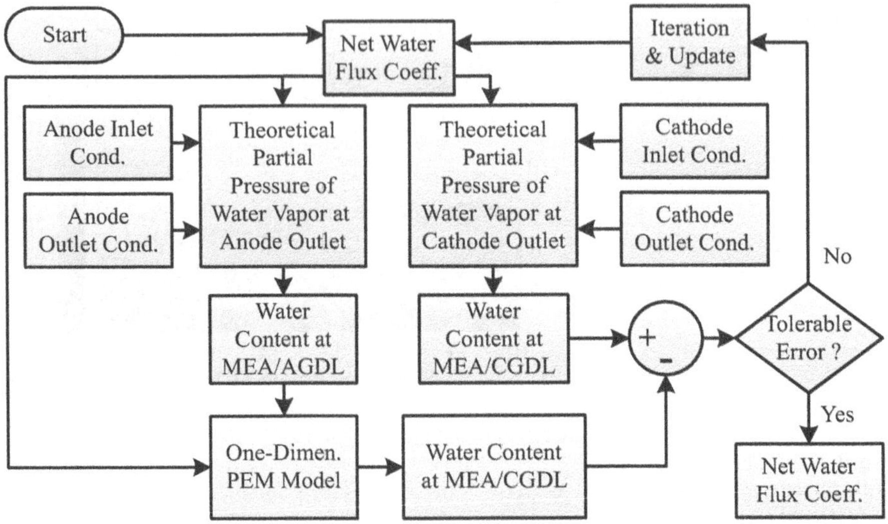

Fig. 4.22 The first method for estimation of net water flux coefficient

flow channel is the arithmetic average or geometric mean of that at the cathode inlet and that at the cathode outlet. Similarly, the concentration of water vapor in the anode flow channel is the arithmetic or geometric average of that at the anode inlet and that at the anode outlet.

Based on the concentration of water vapor in the specific flow channel, the local water content parameter at the interface of the cathode gas diffusion layer and the MEA and that at the interface of the anode gas diffusion layer and the MEA are calculated without considering the water transport in both gas diffusion layers. However, the concentration of water vapor at the ⑤ interface is lower than that at the ⑥ interface (Fig. 2.1), which will results in under-estimated value for water concentration. Because the theoretical concentration of water vapor at cathode outlet is closed related to the net water flux coefficient and the generated water, the concentration of water vapor at the cathode outlet will be more persuasive to be treated as that at the ⑥ interface.

Similarly, the concentration of water vapor at the ⑧ interface is lower than that at the ⑦ interface, which will results in under-estimated value for water concentration. Even so, the concentration of water vapor at the anode outlet will limit the reasonable zone as Sect. 4.4.1. Therefore, the concentration of water vapor at the ⑧ interface will be treated as that at the ⑦ interface.

The estimation of net water flux coefficient is performed as following procedures: firstly, the initial value for the coefficient is set and according to the operating condition at the inlet and outlet of both the cathode and anode, the local water content parameters at both interfaces of the MEA are calculated; secondly, according to the governing equation for water transport in the MEA and the local water content parameter at the interface of the MEA and the anode gas diffusion layer, the local water content parameter at the interface of the MEA and the cathode gas diffusion layer is calculated; thirdly, the local water content parameter at the interface of the MEA and the cathode gas diffusion layer calculated in the first step is compared with that calculated in the second step and we get the error; fourthly, if the error doesn't satisfy the threshold, procedures at the first, second and third step are repeated until the threshold is met.

For example, the operating temperature is 70 °C, the thickness of membrane 45 μm, the stoichiometric ratio of supplied hydrogen 1.5, the relative humidity of supplied hydrogen 50%, the operating pressure of both cathode chamber and anode chamber 150 kPa and the current density is 0.8 and 1.2 A/cm^2. The simulation result is shown (Fig. 4.23) with varied relative humidity and stoichiometric ratio of supplied air.

The net water flux coefficient is smaller than −0.17. For current density at 0.8 A/cm^2, when the air stoichiometric ratio is increased and the relative humidity of supplied air is lower than 20%, the coefficient will be increased slowly. On the contrary, when the stoichiometric ratio of supplied air is increased and the relative humidity of supplied air is larger than 30%, the coefficient will be decreased gradually. When the relative humidity of supplied air ranges from 20 to 30%, the stoichiometric ratio has little influence on the coefficient.

For current density at 1.2 A/cm^2, when the air stoichiometric ratio is increased and the relative humidity of supplied air is lower than 45%, the coefficient will be increased slowly. On the contrary, when the stoichiometric ratio of supplied air is increased and the relative humidity of supplied air is larger than 55%, the coefficient will be decreased gradually. When the relative humidity of supplied air ranges from 20 to 30%, the stoichiometric ratio has little influence on the coefficient. The common conclusion is that the coefficient depends mainly on the current density and the relative humidity of supplied air for certain range of relative humidity of supplied air.

When the relative humidity of supplied air is lower than 10% at 0.8 A/cm^2 current density or that is lower than 35% at 1.2 A/cm^2 current density, the net water flux coefficient cannot be solved based on this method. Qualitatively, when the relative humidity of supplied air is very low, the theoretical partial pressure of water vapor at cathode outlet is lower than the saturated water vapor pressure and the fuel cell tends to be dried out.

Secondly, the estimation of net water flux coefficient for a fuel cell in drying out state is presented (Fig. 4.24). Of course, the partial pressure of water vapor at cathode outlet is higher than that at the cathode flow channel but the over-saturated ratio for water vapor is not that much high. Therefore, it is feasible to use the one-dimensional model of the fuel cell proposed by Springer to calculate the coefficient.

The specific procedures for estimation of water flux coefficient is as follows: firstly, the initial value for the coefficient is set and according to the operating condition at the inlet and outlet of both the cathode and anode, the local water content parameters at both interfaces of the MEA are calculated; secondly, according to the governing equation for the fuel cell and the local water content parameter at the interface of the MEA and the anode gas diffusion layer, the local water content parameter at

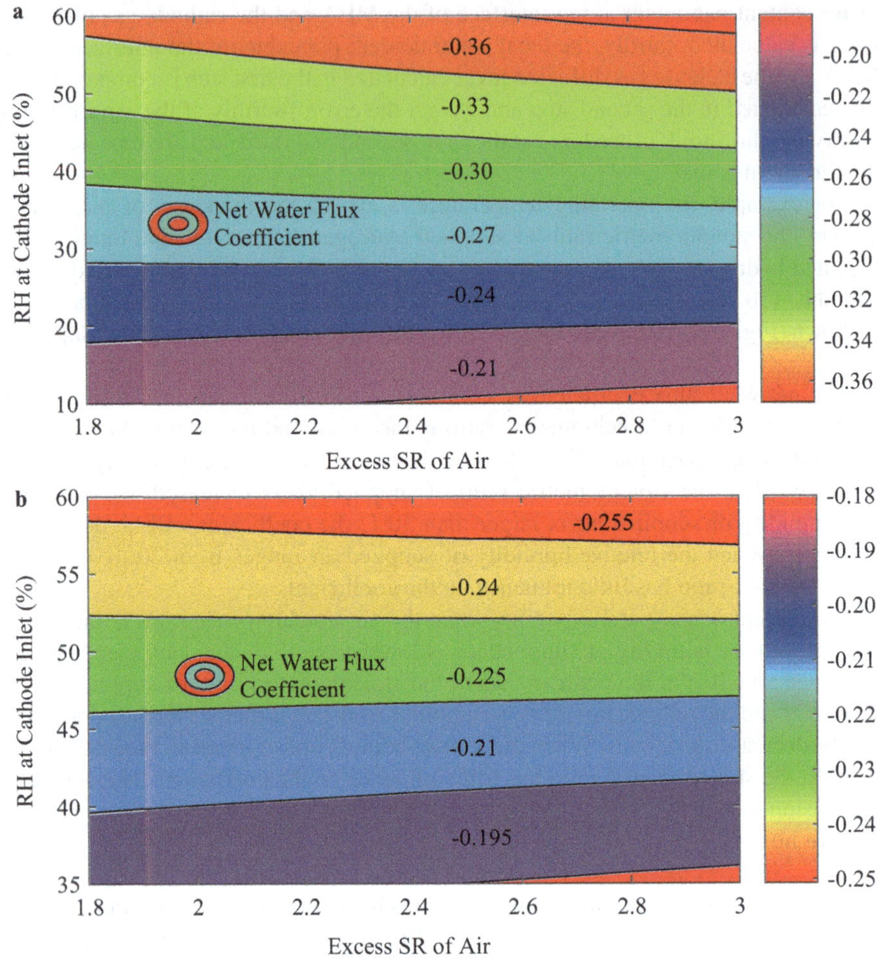

Fig. 4.23 **a** The net water flux coefficient with respect to relative humidity and excess stoichiometric ratio of supplied air at 0.8 A/cm^2, **b** The net water flux coefficient with respect to relative humidity and excess stoichiometric ratio of supplied air at 1.2 A/cm^2

the interface of the MEA and the cathode gas diffusion layer is calculated; thirdly, the local water content parameter at the interface of the MEA and the cathode gas diffusion layer calculated in the first step is compared with that calculated in the second step and we get the error; fourthly, if the error doesn't satisfy the threshold, procedures at the first, second and third step are repeated until the threshold is met. The simulation result won't be elucidated and if interested, please refer to simulation results in Sect. 2.3.

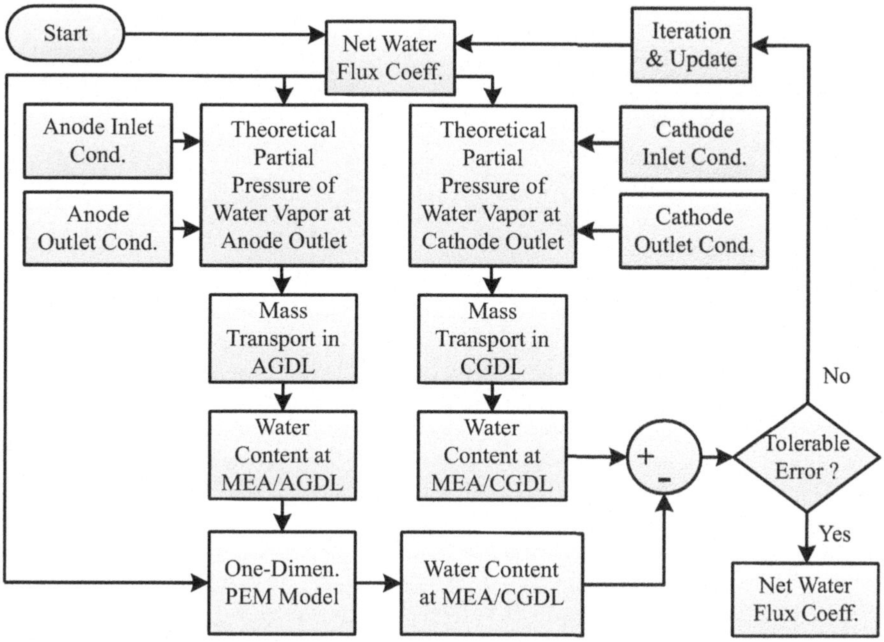

Fig. 4.24 The second method for estimation of net water flux coefficient

4.5 Summary

The method for water content estimation of a fuel cell is proposed in this chapter. Firstly, the error analysis and compensation of the online measured AC impedance is presented and the accuracy of the measured impedance is validated in comparison with that of the electrochemical workstation. Secondly, the high frequency resistance at 400 Hz is the quantitative method for estimating water content in the MEA, the statistical index of the fuel cell voltage in stack is the qualitative method for estimating water content in the MEA and they are feasible for a fuel cell system with self-humidifying MEA. Thirdly, the high frequency resistance together with the method for estimation of water flux coefficient is proposed to calculate the average water content in the MEA with improved accuracy.

Reference

1. Pukrushpan JT, Peng H, Stefanopoulou AG (2004) Control-oriented modeling and analysis for automotive fuel cell systems. J Dyn Syst Measur Control Trans ASME 126(1):14–25

Chapter 5
Water Management Strategy for the PEM Fuel Cell

5.1 Introduction

The high frequency resistance is closely related to water content based on analysis of the MEA model. The approach to online AC impedance measurement acquires the high frequency resistance in real-time. The combination of the high frequency resistance with the statistic index of output voltage of an individual cell and the fuel cell stack contributes to estimating variation trend of the water content like flooding and drying out. The type of MEA suitable for self-humidifying or external-humidifying also affects the online estimation of water content.

The water content dominates performance of the fuel cell system and on basis of the online water content estimation method, the theme of this chapter is to put forward water management strategy. The rest of this chapter is organized as follows: in Sect. 5.2, the control-oriented dynamic model of the fuel cell system is established, including the model of the air supply subsystem, that of the cathode chamber and that of the anode chamber; in Sect. 5.3, the dynamic model of the air supply subsystem is decoupled from the water content estimation method in terms of time constant of their dynamic responses and the control algorithm for the air supply subsystem is proposed; in Sect. 5.4, water management strategy is designed and optimized founded on the characteristic analysis of the fuel cell system with self-humidifying MEA, searching of optimal operating conditions and dynamic experiment of the stack.

5.2 Dynamic Model of the Fuel Cell System

The fuel cell system is composed of four subsystems such as the air supply subsystem, the hydrogen supply subsystem, the cooling subsystem and the control subsystem. The dynamic model of the air supply subsystem is the organic combination of model of parts like the air compressor, humidifier, the cathode chamber and electronic

P. Hong, *Water Content Estimation and Control of PEM Fuel Cell Stack and the Individual Cell in Vehicle*, Springer Theses, https://doi.org/10.1007/978-981-16-8814-0_5

throttle valves. The model of the anode chamber is closed integrated with that of the cathode chamber via net water flux and they will be elucidated.

5.2.1 Dynamic Model of the Cathode Chamber

As Fig. 2.1, the concentration of oxygen, nitrogen, water vapor and liquid water in the cathode chamber varies under different operating conditions. The lumped model for each component is presented as Eq. 5.1, whereas $Q_{Oxy,5}$ and $Q_{W,5}$ are the flux of oxygen and water vapor at the interface of cathode flow channel (CFC) and cathode gas diffusion layer (CGDL), m_{Oxy}, $m_{W,Ca}$, and m_{Nitro} are mass of oxygen, water and nitrogen in the CFC, $Q_{Oxy,in}$, $Q_{W,Ca,In}$ and $Q_{Nitro,\ In}$ are the flux of oxygen, water vapor and nitrogen at the cathode inlet, $Q_{Oxy,Out}$, $Q_{W,Ca,Out}$ and $Q_{Nitro,Out}$ are the flux of oxygen, water vapor and nitrogen at the cathode outlet.

$$\begin{cases} \frac{dm_{Oxy}}{dt} = Q_{Oxy,In} - Q_{Oxy,Out} - Q_{Oxy,5} \\ \frac{dm_{W,Ca}}{dt} = Q_{W,Ca,In} - Q_{W,Ca,Out} + Q_{W,5} \\ \frac{dm_{Nitro}}{dt} = Q_{Nitro,In} - Q_{Nitro,Out} \end{cases} \tag{5.1}$$

$$\begin{cases} T_1 \frac{dQ_{Oxy,6}}{dt} = -Q_{Oxy,6} + Q_{Oxy,5} \\ T_2 \frac{dQ_{Oxy,6}}{dt} = -Q_{Oxy,6} + \frac{I}{4F} M_{Oxy} \end{cases} \tag{5.2}$$

$$\begin{cases} T_3 \frac{dQ_{W,5}}{dt} = -Q_{W,5} + k_{drain} Q_{W,6} \\ T_4 \frac{dQ_{W,6}}{dt} = -Q_{W,6} + \frac{I}{2F} M_W + \beta \frac{I}{F} M_W \end{cases} \tag{5.3}$$

The electrochemical reaction happens in the cathode catalyst layer (CCL) and the consumption rate of oxygen and the generation rate of water are proportional to the output current of the fuel cell. The net water flux from cathode to anode or from anode to cathode depends on current density and operating condition, so the variable $Q_{W,5}$ varies.

In the CCL, the local concentration of oxygen is decreased along the direction of electronic conduction. If the concentration of supplied oxygen is changed, the concentration of oxygen in the CCL will respond with time delay. If the output current of the fuel cell is changed, the concentration of oxygen in the CCL will be changed quickly and the concentration of oxygen in the CFC will respond with time delay as well. Based on this characteristic, dynamic response of the local oxygen flux in the cathode can be described as Eq. 5.2, whereas $Q_{Oxy,6}$ is the oxygen flux at the interface of the CCL and CGDL, M_{Oxy} the mole mass of oxygen, T_1 and T_2 time constant of the dynamic response to each excitation.

On the other hand, the local concentration of water vapor in the CCL is increased along the direction of electronic conduction. If the concentration of water vapor in the supplied air is changed, the concentration of water vapor in the CCL will be respond with time delay. If the output current of the fuel cell is changed, the concentration of water vapor in the CCL will respond very quickly and concentration of water vapor in the CFC will respond with time delay as well. Based on this characteristic, dynamic response of the water flux in the cathode is presented as Eq. 5.3, whereas $Q_{W,6}$ is the water vapor flux at the interface of the CCL and CGDL, M_W the mole mass of water, T_3 and T_4 time constant of the dynamic response to each excitation, k_{drain} quantifies the passive diffusion of water vapor from interface of the CCL and CGDL to interface of the CGDL and CFC, β the net water flux coefficient through the membrane.

$$\int_{t_0}^{t_1} \left(Q_{W,6} - Q_{W,5}\right) dt \leq \left(m_{W,CGDL} + m_{W,CCL}\right)_{LimitH} \tag{5.4}$$

$$\begin{cases} Q_{W,5} = Q_{W,5}\big|_{t_1} + Q_{W,Ca,Step}\left(1 - e^{-\frac{t-t_1}{T_5}}\right), \; t_1 \leq t \leq t_2 \\ \int_{t_1}^{t_3} \left(Q_{W,5} - Q_{W,5}\big|_{t_1}\right) dt \leq \left(m_{W,CGDL} + m_{W,CCL}\right)_{LimitH} \end{cases} \tag{5.5}$$

Dynamics of the water vapor and droplet in the cathode is presented (Fig. 5.1). If water vapor at the cathode outlet is saturated, liquid water emerges in the form of water film on surface of the CGDL and in the form of water droplet in the CFC. The droplet is formed mainly when the liquid water content on surface of the CGDL exceeds the critical value which depends on the air flowrate, temperature, contact angle and water storage capability of the CGDL. To some extent, the water film is

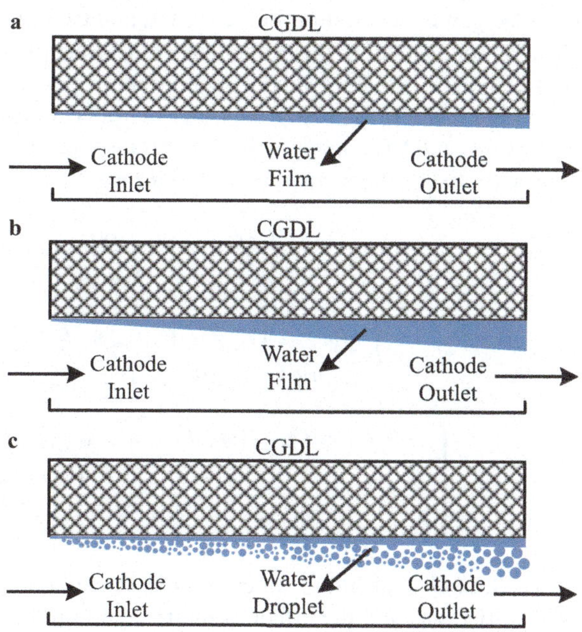

Fig. 5.1 Dynamic of liquid water content in the cathode flow channel: **a** initial accumulation of liquid water as film, **b** the critical content of liquid water as film, **c** formation of water droplet

broken up by the air flow rate and then the droplet emerges. The critical value is presented as Eq. 5.4 and once the integration of difference between $Q_{W,6}$ and $Q_{W,5}$ with time is close to the critical value, the possibility of formation of water droplet is increased.

Both water vapor, water film and water droplet are carried away by the air flow but the movement of water is dynamic, of which the direct indicator is the dynamic of the gas pressure at the cathode outlet. It results from continuous slow accumulation of liquid water in the CCL and CGDL, continuous removal of water vapor in the CFC but intermittent fast drainage of water droplet on surface of the CGDL. Therefore, the coefficient k_{drain} is smaller than 1. The fast drainage of water droplet on surface of the CGDL promotes the transient step response $Q_{W,Ca,Step}$ of the water flux at the interface of the CGDL and CFC, as Eq. 5.5, whereas t_0, t_1, t_2 and t_3 are the moment when the dynamic happens, T_5 the time constant for removal of water droplet. The pressure at the cathode outlet will be decreased for short time and then it recovers. Once the water droplet is removed, the dynamic of liquid water in the CFL will be repeated again.

The difference between dynamics of oxygen and water lies in the driven force of mass transport. To be specific, water transport is driven only by the gradient of local water concentration in the CGDL and CCL and only when the local water concentration in the CFC is lower than that in the CGDL or CCL, the water can be transported. That is to say, the dynamic of water transport in the CFC determines that in the CGDL or CCL. The combination of water vapor, water film and droplet makes this dynamic complicated and slow. The boundary condition for the model of water is subject to the Robin boundary condition in the CFC, CGDL and CCL. On the contrary, even though the oxygen transport is also driven only by the gradient of local oxygen concentration in the CGDL and CCL, the oxygen consumption in the CCL is proportional to the output current and the dynamic of oxygen in the CCL is subject to the Neumann boundary condition for gas transport. Therefore, the dynamic of oxygen transport is faster than that of water transport and the oxygen transport in the CGL and CCL hinders the water transport due to counter direction. The time constant satisfies the inequality as $T_2 < T_1 << T_4 < T_3$.

$$\begin{cases} Q_{LW} = k_{Liq} Q_{Gas,Ca,Out} \\ Q_{GW} + Q_{LW} \leq Q_{W,5} \end{cases} \tag{5.6}$$

$$\begin{cases} Q_{W,5} = Q_{W,6} = (1+2\beta)\frac{I}{2F}M_W - \frac{dm_{W,MEA}}{dt} \\ m_{W,MEA} \geq \left(m_{W,MEA}\right)_{LimitL} = m_0(I) \end{cases} \tag{5.7}$$

$$\begin{cases} \frac{dm_{W,MEA}}{dt} = k_{MEA}\left(m_0(I) - m_{W,MEA}\right)\frac{Q_{Ca,Out}}{Q_0} f\left(P_{W,Ca,Out}\right) \\ P_{sat}\left(T_{Ca,Out}\right) \geq P_{W,Ca,Out} \\ f\left(P_{W,Ca,Out}\right) = \frac{P_{sat}\left(T_{Ca,Out}\right) - P_{W,Ca,Out}}{P_{sat}\left(T_{Ca,Out}\right)} \end{cases} \tag{5.8}$$

The flux Q_{LW} of liquid water at the cathode outlet is function of flux $Q_{Gas,Ca,Out}$ of the dried gas and the effect of temperature on the flux Q_{LW} is quantified by the coefficient k_{Liq}, as Eq. 5.6. Of course, sum of the flux Q_{LW} of liquid water and the flux Q_{GW} of water vapor at the cathode outlet is smaller than the flux of water at the interface of CGDL and CFC.

If water vapor at the cathode outlet is not saturated, the water vapor flux at the interface of the CGDL and CFC is the same as that at the interface of the CGDL and CCL and liquid water may still accumulate in the CGDL and CCL but with no significant impact, as Eq. 5.7. The pressure at the cathode outlet won't experience intermittent decrease and recovery. The dried air flux in the CFC is capable of carrying more water from the CGDL and CCL and water content $m_{W,MEA}$ in the MEA will be decreased gradually. On condition that the fuel cell is still working, the water content in the MEA won't be decreased to zero and the lower limit $m_0(I)$ is function of the output current and operating condition. The variation of water content in the MEA with time is function of its current water content, air flux $Q_{Ca,Out}$ and the partial pressure $P_{W,Ca,Out}$ of water vapor at the cathode outlet, as Eq. 5.8. The partial pressure of water vapor is lower than the saturated pressure P_{sat} corresponding to the temperature $T_{Ca,Out}$ at the cathode outlet. The coefficient k_{MEA} is introduced to take the temperature into consideration and Q_0 is the reference air flux.

If the operating condition of supplied air is changed and the water vapor at the cathode outlet becomes saturated from unsaturated, the water flux at the interface of the CGDL and CFC and at the interface of the CCL and CGDL is decreased significantly within short time. The accumulation of liquid water in the CFC, CGDL and CCL is enhanced and water content in the MEA begins to increase at the same time. During this time, the HFR is decreased but with time delay T_6.

If the operating condition of supplied air is changed and the water vapor at the cathode outlet becomes unsaturated from saturated, liquid water in the CFL will be removed quickly and then the water in the CGDL is decreased. Finally, water content in the CCL and membrane is reduced. During this time, the HFR is increased but with time delay T_7.

Taking all dynamics of the cathode chamber into consideration, the times constant satisfies the inequality as $T_2 < T_1 << T_5 < T_4 < T_3 << T_6 < T_7$. The model of cathode chamber of the fuel cell stack is the lumped model of cathode chamber of each cell and if the manifold for gas distribution to each cell is designed to guarantee even gas flow rate and pressure, the lumped model can be simplified as the average model. If not, the lumped model will be complex.

5.2.2 *Dynamic Model of the Anode Chamber*

As Fig. 2.1, gas components in the anode chamber changes slowly with operating condition because the flow rate of hydrogen and water vapor is significantly smaller than that of air and water vapor in the cathode chamber. For example, the mole fraction of oxygen in air is around 21% and if the excess stoichiometric ratio of oxygen is 2 and that of hydrogen 2 as well, the flow rate at the cathode outlet is almost five times of that at the anode outlet. According to the mass conservation, the mass m_{Hy} of hydrogen and the mass $m_{W,An}$ of water in the anode chamber are presented as Eq. 5.9, whereas M_{Hy} is mole mass of hydrogen, $Q_{Hy,In}$ and $Q_{W,An,In}$ the flux of hydrogen and water vapor at the anode inlet, $Q_{Hy,Out}$ and $Q_{W,An,Out}$ the flux of hydrogen and water at the anode outlet, and $Q_{W,An,8}$ the flux of water at the interface of the anode gas diffusion layer (AGDL) and anode flow channel (AFC).

The transport of hydrogen in the AGDL and the anode catalyst layer (ACL) is driven by the electrochemical reaction consumption and it's proportional to the output current. The boundary condition for flux of hydrogen transport in the anode chamber is subject to the Neumann boundary condition. The inter-diffusion between hydrogen and water vapor is larger than that between oxygen and water vapor. The water flux through the AGDL and ACL is significantly smaller than that through the CGDL and CCL, and it's maybe 10% of the generated water. Except for the operating condition resulting in starvation of hydrogen, the dynamic of hydrogen in the AGDL and ACL can be neglected in comparison with that of water in the AGDL and ACL.

$$\begin{cases} \frac{dm_{Hy}}{dt} = Q_{Hy,In} - Q_{Hy,Out} - \frac{I}{2F} M_{Hy} \\ \frac{dm_{W,An}}{dt} = Q_{W,An,In} - Q_{W,An,Out} - Q_{W,An,8} \end{cases} \tag{5.9}$$

$$\begin{cases} \frac{dm_{W,AGDL}}{dt} = -\beta \frac{I}{F} M_W - Q_{W,An,8} \\ T_8 \frac{dQ_{W,An,8}}{dt} = -Q_{W,An,8} + k_{drain,An} \beta \frac{I}{F} M_W \end{cases} \tag{5.10}$$

$$\begin{cases} Q_{W,8} = Q_{W,8}\big|_{t_4} + Q_{W,An,Step}\left(1 - e^{-\frac{t-t_4}{T_9}}\right), \quad t_4 \le t \le t_5 \\ \int_{t_4}^{t_6} \left(Q_{W,8} - Q_{W,8}\big|_{t_4}\right) dt \le \left(m_{W,AGDL}\right)_{LimitH} \end{cases} \tag{5.11}$$

For the fuel cell system, the hydrogen is stored in the hydrogen tank with high pressure and along the path from the hydrogen tank to the anode chamber of the stack, no humidifier is used to humidifying the hydrogen. However, the hydrogen recirculation pump, installed between the anode inlet and outlet, provides exhausted humidified hydrogen to the anode inlet, such that the hydrogen from the hydrogen tank is still humidified in way of mixture. In terms of the anode subsystem, the water originates from the net water flux from the cathode chamber to the anode chamber through the membrane. Accumulation of water vapor and liquid water in the anode chamber leads to decrease of concentration of hydrogen in the anode chamber if the gas pressure at the anode inlet keeps constant.

The dynamic of water in the AGDL is presented as Eq. 5.10, whereas $k_{drain,An}$ describes the capability of carrying water away through hydrogen flow. The water transport in the AGDL and ACL is governed by the Robin boundary condition and the gradient of water concentration between the AGDL, ACL and AFC drives the water transport. When liquid water accumulates in the ACL due to net water flux through the membrane, the water flux at the interface of the ACL and AGDL responds with time delay T_8. The purge valve installed close to the anode outlet is opened to exhaust gas components in the anode chamber and there will be a step excitation $Q_{W,An,Step}$ of water removing to the AFC and the interface of the AFC and AGDL, as Eq. 5.11, whereas t_4 and t_5 describes when the step excitation starts and ends respectively, and T_9 the time constant. The purge action promotes the water removal and water content in the AGDL will be decreased, but there exists the limit of total removed water content in the anode chamber before next purge action. The time constant satisfies the inequality as $T_9 < T_8$.

If the total amount of water content in the fuel cell is decreased gradually, the total amount of water content in the anode chamber will be decreased as well, but the structure of the anode subsystem guarantees that there will always be water in the anode chamber. In addition, the model of the anode chamber of the fuel cell stack is the lumped model of the anode chamber of each cell and if the manifold for gas distribution to each cell is designed to guarantee even gas flow rate and pressure, the lumped model can be simplified as the average model. If not, the lumped model will be complex.

5.2.3 Dynamic Model of the Air Supply Subsystem

The typical scheme of the air supply subsystem of the fuel cell is presented (Fig. 5.2). To be specific, the air compressor provides pressurized air with high flow rate and the air temperature at the compressor outlet is usually higher than the required at the cathode inlet. The inter-cooling apparatus cools down the pressurized air by providing coolant with lower temperature. The forward electronic throttle valve (ETVF) keeps widely opened when the air supply subsystem starts to work and the open angle of the

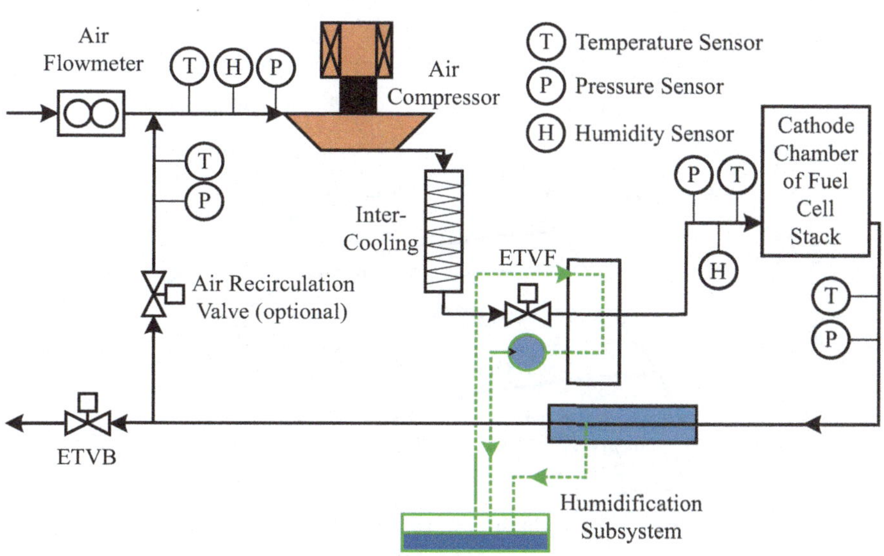

Fig. 5.2 Scheme of the air supply subsystem

back electronic throttle valve (ETVB) is adjusted to ensure proper gas pressure at the cathode inlet. After the fuel cell system is stopped, the ETVF and ETVB are tightly closed to prevent oxygen from penetrating into the cathode chamber. The humidification subsystem recycles part of water leaving the cathode chamber to humidify the pressurized air into the cathode chamber. The temperature sensor, pressure sensor and even humidity sensor are selectively installed for system application. In some cases, the air recirculation valve or pump is installed to lower the oxygen concentration to the cathode chamber to clamp output voltage of the fuel cell at low load for durability.

The air compressor is composed of the high speed motor, transmission and the compressor [1]. The compressor is mechanically connected to the rotor of the high speed motor through the transmission. The high speed motor converts electrical energy to mechanical energy and the rotor together with the compressor spins. The compressor compresses the air and converts mechanical energy to the internal energy and mechanical energy of air. Of course, energy loss exists through the whole chain of energy conversion. The typical MAP of the air compressor is presented (Fig. 5.3) with respect to the pressure ratio, mass flow and compressor speed as well as efficiency. The pressure ratio is the air pressure at the compressor outlet divided by the air pressure at the compressor inlet.

For each constant rotation speed of the compressor, the pressure ratio decreases with increased air mass flow. The operating point with maximal pressure ratio and the minimal air mass flow for each constant rotation speed is called the surge point and connecting all these points together forms the surge line. The operating point with minimal pressure ratio and the maximal air mass flow for each constant rotation speed is called the choke point and connecting all these point together forms the choke line.

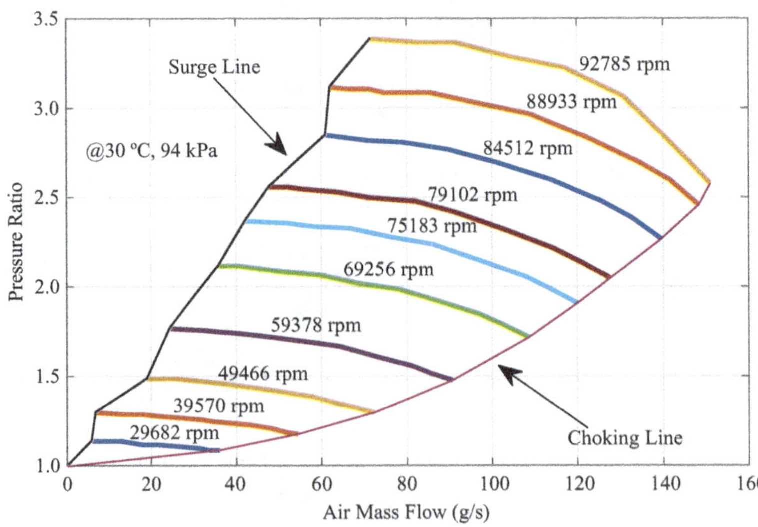

Fig. 5.3 Typical MAP of the air compressor

In the two-dimensional map, only the zone between the surge line and the choke line is feasible and it will damage the compressor permanently if operated beyond this zone. Please refer to the paper [2] for more information about data processing of the MAP of the air compressor.

$$J_{cp}\frac{d\omega_{cp}}{dt} = \tau_{cm} - \tau_{cp} - f_{cp}\omega_{cp} \tag{5.12}$$

$$\tau_{cm} = \eta_{cm}\frac{k_t}{R_{cm}}(v_{cm} - k_v\omega_{cp}) \tag{5.13}$$

$$\tau_{cp} = \frac{C_p T_{atm}}{\omega_{cp}\eta_{cp}}\left[\left(\frac{P_{PipeUp}}{P_{atm}}\right)^{(\gamma-1)/\gamma} - 1\right]W_{cp} \tag{5.14}$$

$$T_{cp} = T_{atm} + \frac{T_{atm}}{\eta_{cp}}\left[\left(\frac{P_{PipeUp}}{P_{atm}}\right)^{(\gamma-1)/\gamma} - 1\right] \tag{5.15}$$

The mechanical dynamic of the air compressor is described as Eqs. 5.12–5.13. Whereas, J_{cp} is sum of moment of inertia of the motor rotor, equivalent moment of inertia of the transmission shaft and gear in view of the motor rotor, and equivalent moment of inertia of the compressor in view of the motor rotor; τ_{cp} is the load torque; ω_{cp} is the rotation speed of rotor; $f_{cp}\omega_{cp}$ is the friction torque and f_{cp} the friction coefficient; τ_{cm} is drive torque of the motor and it's related to power supply v_{cm} of the motor, the rotation speed, the structural parameter k_t and R_{cm}; η_{cm} is energy conversion efficiency of the motor.

The aerodynamics of the air compressor is present as Eqs. 5.14–5.15. Whereas, W_{cp} is the air mass flow, η_{cp} the compression efficiency, T_{cp} and P_{PipeUp} the temperature and pressure of air at the compressor inlet, T_{atm} and P_{atm} the temperature and pressure of air at the compressor outlet, C_p the specific heat capacity of air at constant pressure, γ the specific heat capacity ratio of air.

$$\frac{dm_{PipeUp}}{dt} = W_{cp} - W_{PipeUp,Out} \tag{5.16}$$

$$\frac{dP_{PipeUp}}{dt} = \frac{\gamma R_{Air}}{V_{PipeUp}}\left(W_{cp}T_{cp} - W_{PipeUp,Out}T_{PipeUp}\right) \tag{5.17}$$

$$W_{Ca,In} = W_{PipeUp,Out} = k_{PipeUpCa}\left(P_{PipeUp} - P_{Ca}\right) \tag{5.18}$$

$$W_{Ca,Out} = W_{PipeDS,In} = k_{CaPipeDS}\left(P_{Ca} - P_{PipeDS}\right) \tag{5.19}$$

$$\frac{dP_{PipeDS}}{dt} = \frac{R_{Air}T_{PipeDS}}{V_{PipeDS}}\left(W_{Ca,Out} - W_{PipeDS,Out}\right) \tag{5.20}$$

$$W_{PipeDS,Out} = \frac{C_{Thrt}A_{Thrt}P_{PipeDS}}{\sqrt{RT_{PipeDS}}}\left(\frac{P_{atm}}{P_{PipeDS}}\right)^{\frac{1}{\gamma}}\sqrt{\frac{2\gamma}{\gamma-1}\left(1-\left(\frac{P_{atm}}{P_{PipeDS}}\right)^{\frac{\gamma-1}{\gamma}}\right)} \tag{5.21}$$

$$A_{Thrt} = \frac{\pi d_{ETV}^2}{4}(1-\cos\theta_{ETV}) \tag{5.22}$$

The pipe between the compressor outlet and the cathode inlet is considered as the first lumped chamber of which the volume is constant as V_{PipeUp}. In this lumped chamber, the temperature, pressure and gas component are considered to be uniform. According to mass conservation and energy conservation, the mass m_{PipeUp} and pressure P_{PipeUp} of air in the lumped chamber are described as Eqs. 5.16–5.17, whereas $W_{PipeUp,Out}$ is the air mass flow leaving the lumped chamber, R_{Air} the gas constant of air, and T_{PipeUp} the temperature of air in the lumped chamber [1]. The dynamic of the humidification subsystem and inter-cooling apparatus will not be considered in our research.

The air leaving the first lumped chamber enters the cathode inlet and due to small difference between pressure P_{PipeUp} of the first lumped chamber and pressure P_{Ca} of the cathode chamber, the air mass flow $W_{Ca,Out}$ and $W_{PipeUp,Out}$ at this interface are presented as Eq. 5.18, whereas $k_{PipeUpCa}$ is the air flow coefficient.

The pipe between the cathode outlet and the upstream of the ETVB is considered as the second lumped chamber of which the volume is constant as V_{PipeDS}. The air leaving the cathode outlet enters the second lumped chamber and due to small difference between pressure P_{PipeDS} of the second lumped chamber and pressure P_{Ca} of the cathode chamber, the air mass flow $W_{Ca,Out}$ and $W_{PipeDS,In}$ at the interface are presented as Eq. 5.19, whereas $k_{CaPipeDS}$ is the air flow coefficient. The temperature, pressure and gas component are assumed to be the same anywhere of the second lumped chamber. According to mass conservation, the pressure P_{PipeDS} of the second lumped chamber is governed by Eq. 5.20.

If the open angle of the ETVB is decreased, the effective section area for mass flow is reduced and the pressure drop through the ETVB increased. On the other hand, if the open angle of the ETVB is increased, the effective section area for mass flow is increased and the pressure drop through the ETVB decreased. In this way, the pressure at the cathode chamber is changed through adjustment of the open angle of the ETVB and the operating point of the air compressor changed. The downstream of the ETVB is the environment and the relation between mass flow W_{PipeDS} through the ETVB and the pressure P_{PipeDS} of the second lumped chamber is presented as Eq. 5.21. The effective section area A_{Thrt} for mass flow is presented as Eq. 5.22, whereas C_{Thrt} is the orifice coefficient, d_{ETV} and θ_{ETV} the diameter and open angle of the ETVB respectively.

5.3 Control Algorithm of the Air Supply Subsystem

To start with, time constant of the air supply subsystem is analyzed. Based on the multi-component Stephan-Maxwell diffusion equation, the time constant for oxygen transport in the CGDL is around 0.013 s, if the fuel cell is operated at 60 °C, the thickness and porosity of the CGDL 250 μm and 0.3 respectively. The time constant for dynamic of the air compressor is around 0.8 s based on data by Kingston Corporation. As described in Sects. 4.3 and 5.2.1, the time constant for water content of the fuel cell is as long as tens of seconds and it depends on the operating condition. The time constant for the ETV by Continental Corporation is about 0.1 s. The five mentioned parts have different time scales and it is suggested to design control algorithm based on divided frequency domain.

The target for control algorithm of the air supply subsystem is gas pressure at the cathode inlet and air mass flow into the compressor. The state variable $\boldsymbol{x}$ is gas pressure of the cathode chamber, rotation speed of the compressor, gas pressure of the first lumped chamber and gas pressure of the second lumped chamber, as Eq. 5.23. The input variable $\boldsymbol{u}$ is voltage supply of the compressor, output current of the fuel cell and open angle of the ETVB, as Eq. 5.24. The measurable variables are air mass flow into the compressor, temperature of environment air, gas temperature and pressure at the compressor outlet, gas temperature and pressure at the cathode inlet, output voltage of the fuel cell, rotation speed and power consumption of the compressor and so on. The selected observable variable $\boldsymbol{y}$ is gas pressure of the cathode chamber, rotation speed of the compressor and gas pressure of the first lumped chamber, as Eq. 5.25. Again in this section, the dynamic of gas temperature in the first lumped chamber won't be considered and the gas temperature at the cathode inlet is assumed to satisfy system requirement all the time.

Flow chart of the control algorithm of the air supply subsystem is presented (Fig. 5.4). Firstly, the error between target variables and measured variables is input to the closed-loop control algorithm of the air compressor, the first lumped chamber,

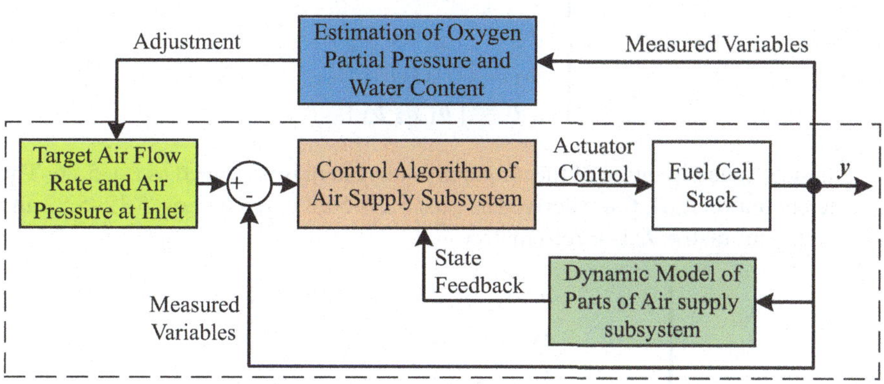

Fig. 5.4 Control algorithm of the air supply subsystem

pressure of the cathode chamber, the ETVB and the second lumped chamber. Secondly, the output of the closed-loop control algorithm, including rotation speed of the compressor and open angle of the ETVB, is applied to the corresponding actuator of the fuel cell system. Thirdly, based on state estimation of part of the air supply subsystem, the closed-loop control algorithm is optimized. Fourthly, part of measured variables are applied to estimate oxygen partial pressure and water content in the cathode chamber. Finally, based on estimation result of the oxygen and water content, target variables are adjusted to optimize performance of the fuel cell with appropriate water content.

$$x = \left[P_{Ca}\ \omega_{cp}\ P_{PipeUp}\ P_{PipeDS} \right]^T \tag{5.23}$$

$$\boldsymbol{u} = \left[v_{cm}\ I\ \theta_{ETC} \right]^T \tag{5.24}$$

$$y = \left[P_{Ca}\ W_{cp}\ P_{PipeUp} \right]^T \tag{5.25}$$

$$\begin{cases} \dot{x} = \boldsymbol{Ax} + \boldsymbol{Bu} \\ y = \boldsymbol{Cx} + \boldsymbol{Du} \end{cases} \tag{5.26}$$

$$\boldsymbol{A} = \begin{bmatrix} A_{11} & 0 & A_{13} & A_{14} \\ 0 & A_{22} & A_{23} & 0 \\ A_{31} & A_{32} & A_{33} & 0 \\ A_{41} & 0 & 0 & A_{44} \end{bmatrix} \tag{5.27}$$

$$\boldsymbol{C} = \begin{bmatrix} 1 & 0 & 0 & 0 \\ 0 & k_{Wcp1} & k_{Wcp2} & 0 \\ 0 & 0 & 1 & 0 \end{bmatrix} \tag{5.28}$$

$$\begin{cases} \vec{\boldsymbol{b}}_1 = \left[0\ \frac{\eta_{cm} k_t}{R_{cm} J_{cp}}\ 0\ 0 \right]^T \\ \vec{\boldsymbol{b}}_2 = \left[b_{21}\ 0\ 0\ 0 \right]^T \\ \vec{\boldsymbol{b}}_3 = \left[0\ 0\ 0\ b_{33} \right]^T \\ \boldsymbol{B} = \left[\vec{\boldsymbol{b}}_1\ \vec{\boldsymbol{b}}_2\ \vec{\boldsymbol{b}}_3 \right] \end{cases} \tag{5.29}$$

The state space equation of the air supply subsystem is presented as Eq. 5.26. The coefficient matrix $\boldsymbol{A}, \boldsymbol{B}, \boldsymbol{C}$ are derived as Eqs. 5.27–5.38 based on Eqs. 5.12–5.22 and the coefficient matrix $\boldsymbol{D}$ is a zero matrix.

$$\begin{cases} A_{11} = -\frac{R_{Air} T_{Ca}^{*} k_{PipeUpCa}}{V_{Ca}} - \frac{R_{AirRH} T_{Ca}^{*} k_{PipeUpCa}}{V_{Ca}} \\ A_{13} = k_{PipeUpCa} \frac{R_{air} T_{Ca}^{*}}{V_{Ca}} \\ A_{14} = k_{PipeUpCa} \frac{R_{Air} T_{Ca}^{*}}{V_{Ca}} \end{cases} \tag{5.30}$$

$$A_{22} = -\frac{\eta_{cm} k_t k_v}{R_{cm} J_{cp}} - \frac{f_{cp}}{J_{cp}} + \frac{c_p T^*_{atm}}{\left(\omega^*_{cp}\right)^2 \eta_{cp} J_{cp}} \left[\left(\frac{P^*_{PipeUp}}{P^*_{atm}}\right)^{(\gamma-1)/\gamma} - 1\right] W^*_{cp} \tag{5.31}$$

$$A_{23} = -\frac{(\gamma-1) c_p T^*_{atm} W^*_{cp}}{\gamma \omega^*_{cp} \eta_{cp} J_{cp}} \left(P^*_{PipeUp}\right)^{-1/\gamma} \left(P^*_{atm}\right)^{(1-\gamma)/\gamma} \tag{5.32}$$

$$\begin{cases} A_{31} = RT^*_{cp} k_{PipeUpCa} / \left(V_{PipeUp} M_{air}\right) \\ A_{32} = RT^*_{cp} k_{Wcp1} / \left(V_{PipeUp} M_{air}\right) \end{cases} \tag{5.33}$$

$$A_{41} = \frac{R_{Air} T^*_{PipeDS}}{V_{PipeDS}} k_{CaPipeDS} \tag{5.34}$$

$$\begin{cases} g_1 = RT^*_{cp}\left(k_{Wcp2} - k_{PipeUpCa}\right) / \left(V_{PipeUp} M_{air}\right) \\ g_2 = \frac{R\left[W^*_{cp} + \left(P^*_{Ca} - P^*_{PipeUp}\right) k_{PipeUpCa}\right]}{V_{PipeUp} M_{air}} \\ A_{33} = g_1 + g_2 \frac{(\gamma-1) T^*_{atm}}{\gamma \eta_{cp}} \left(P^*_{PipeUp}\right)^{-1/\gamma} \left(P^*_{atm}\right)^{(1-\gamma)/\gamma} \end{cases} \tag{5.35}$$

$$\begin{cases} A_{44} = \frac{R_{Air} T^*_{PipeDS}}{V_{PipeDS}} \left(-k_{CaPipeDS} - g_3\right) \\ g_3 = \left.\frac{\partial W_{PipeDS,Out}}{\partial P_{PipeDS}}\right|_{SteadyState} \end{cases} \tag{5.36}$$

$$b_{31} = \frac{N_{stack} RT^*_{Ca} (1 + 4\beta)}{4 V_{Ca} F} \tag{5.37}$$

$$b_{33} = -\frac{R_{Air} T^*_{PipeDS}}{V_{PipeDS}} \left.\frac{\partial W_{PipeDS,Out}}{\partial \theta_{ETC}}\right|_{SteadyState} \tag{5.38}$$

The air supply subsystem is constrained as Eqs. 5.39–5.41. To be specific, four variables have both upper and lower limit, namely voltage supply of the compressor, output current of the fuel cell, open angle of the ETVB, and gas pressure of the cathode chamber. Dynamic variation of these variables is limited as well for system stability. For each constant rotation speed of the air compressor, the air mass flow into the compressor is limited between the surge point and choke point.

$$\begin{cases} (v_{cm})_{\min} \le v_{cm} \le (v_{cm})_{\max} \\ (\Delta v_{cm})_{\min} \le \Delta v_{cm} \le (\Delta v_{cm})_{\max} \\ I_{\min} \le I \le I_{\max} \\ \Delta I_{\min} \le \Delta I \le \Delta I_{\max} \end{cases} \tag{5.39}$$

$$\begin{cases} (\theta_{ETC})_{\min} \le \theta_{ETC} \le (\theta_{ETC})_{\max} \\ (\Delta \theta_{ETC})_{\min} \le \Delta \theta_{ETC} \le (\Delta \theta_{ETC})_{\max} \\ (P_{Ca})_{\min} \le P_{Ca} \le (P_{Ca})_{\max} \\ (\Delta P_{Ca})_{\min} \le \Delta P_{Ca} \le (\Delta P_{Ca})_{\max} \end{cases} \tag{5.40}$$

$$\begin{cases} (W_{cp})_{MAP} = f(\omega_{cp}, P_{PipeUp}) \\ (W_{cp})_{\min} \le W_{cp} \le (W_{cp})_{\max} \quad for \quad \omega_{cp} = const \end{cases} \tag{5.41}$$

The continuous system needs to be discretized for practical application and sample period of the micro controller unit is T_s. The discretized state space equation of the subsystem is derived as Eq. 5.42. The initial moment is noted as t_0 and the variable $\boldsymbol{x}(k)$ means the state variable at time $kT_s + t_0$. To follow variation of the input variable $\boldsymbol{u}$, the state variable $\boldsymbol{x}(k)$ is augmented with the variable $\boldsymbol{x}_u(k) = \boldsymbol{u}(k-1)$, as Eq. 5.43. The initial state variable is $\boldsymbol{x}_0 = \boldsymbol{x}(t_0)$ and the initial input variable $\boldsymbol{u}_0 = \boldsymbol{u}(t_0)$. For system with multi-constraint, the model predictive control algorithm (MPC) is recommended and the cost function is presented as Eq. 5.44, whereas N is the predicted step, $\boldsymbol{Q}$ and $\boldsymbol{R}$ the weighting matrix, and $\boldsymbol{r}^*$ the target variable.

$$\begin{cases} \boldsymbol{x}(k+1) = \boldsymbol{A}_k\boldsymbol{x}(k) + \boldsymbol{B}_k\boldsymbol{u}(k) \\ \boldsymbol{y}(k) = \boldsymbol{C}_k\boldsymbol{x}(k) + \boldsymbol{D}_k\boldsymbol{u}(k) \end{cases} \tag{5.42}$$

$$\begin{cases} \begin{bmatrix} \boldsymbol{x}(k+1) \\ \boldsymbol{x}_u(k+1) \end{bmatrix} = \begin{bmatrix} \boldsymbol{A}_k & \boldsymbol{B}_k \\ \vec{\boldsymbol{0}} & \boldsymbol{I} \end{bmatrix} \begin{bmatrix} \boldsymbol{x}(k) \\ \boldsymbol{x}_u(k) \end{bmatrix} + \begin{bmatrix} \boldsymbol{B}_k \\ \boldsymbol{I} \end{bmatrix} \Delta\boldsymbol{u}_k \\ \boldsymbol{y}(k) = \begin{bmatrix} \boldsymbol{C}_k & \vec{\boldsymbol{0}} \end{bmatrix} \begin{bmatrix} \boldsymbol{x}(k) \\ \boldsymbol{x}_u(k) \end{bmatrix} + \boldsymbol{D}_k\Delta\boldsymbol{u}_k \end{cases} \tag{5.43}$$

$$J = \sum_{k=0}^{N-1} \left[\left(\boldsymbol{y}(k) - \boldsymbol{r}^*\right)^T \boldsymbol{Q} \left(\boldsymbol{y}(k) - \boldsymbol{r}^*\right) + (\Delta\boldsymbol{u}(k))^T \boldsymbol{R} \Delta\boldsymbol{u}(k) \right] \tag{5.44}$$

$$\min_{\Delta \boldsymbol{U}} J(\Delta\boldsymbol{U}, \boldsymbol{X}_0) \tag{5.45}$$

$$\frac{\partial J(\Delta\boldsymbol{U}, \boldsymbol{X}_0)}{\partial(\Delta\boldsymbol{U})} = \boldsymbol{H}\Delta\boldsymbol{U} + \begin{bmatrix} \boldsymbol{X}_0^T & (\boldsymbol{r}^*)^T \end{bmatrix} \boldsymbol{F} = 0 \tag{5.46}$$

$$\begin{cases} \Delta\boldsymbol{U} = \begin{bmatrix} (\Delta\boldsymbol{u}(0))^T & (\Delta\boldsymbol{u}(1))^T & \ldots & (\Delta\boldsymbol{u}(N-1))^T \end{bmatrix}^T \\ \boldsymbol{X}_0 = \begin{bmatrix} \boldsymbol{x}_0 \\ \boldsymbol{u}_0 \end{bmatrix} \end{cases} \tag{5.47}$$

Optimized solution to the control algorithm ensures minimal of the cost function with time, as Eqs. 5.45–5.46, whereas $\boldsymbol{H}$ and $\boldsymbol{F}$ are coefficient matrices. The derived solution is in the form as Eq. 5.47 but the procedure won't be elucidated here.

5.4 Water Management Strategy for the Fuel Cell System

In control algorithm of the air supply subsystem, the closed-loop control of actuators to fulfill the target of gas pressure in the cathode chamber and air mass flow into the cathode chamber is proposed. The estimation of oxygen partial pressure and water content in the cathode chamber will be analyzed as follows.

5.4.1 The Fuel Cell System with no External Humidifier

For the fuel cell system with no external humidifier, the relative humidity $RH_{Ca,In}$ at the cathode inlet is determined by pressure P_{atm}, temperature T_{atm} and relative humidity RH_{atm} of the environment, pressure $P_{Ca,In}$ and temperature $T_{Ca,In}$ at the cathode inlet, as Eq. 5.48. In practical system, pressure $P_{Ca,Out}$ and temperature $T_{Ca,Out}$ at the cathode outlet are different from those at the cathode inlet.

$$RH_{Ca,In} = P_{Ca,In}\frac{RH_{atm}P_{sat}(T_{atm})}{P_{atm}P_{sat}\left(T_{Ca,In}\right)} \tag{5.48}$$

$$Q_{Air,In} = Q_{DryAir,In} + Q_{Vapor,In} \tag{5.49}$$

$$\begin{cases} Q_{DryAir,In} = N_{stack}\lambda_{Oxy}\frac{I}{4F}M_{Air} \\ Q_{Air,In} = \frac{Q_{DryAir,In}x_{W,atm}M_{W}}{M_{Air}\left(1-x_{W,atm}\right)} + Q_{DryAir,In} \end{cases} \tag{5.50}$$

$$P_{W,Ca,Out} = P_{Ca,Out}\frac{x_{W,atm} + (1+4\beta)\frac{N_{stack}I}{4FQ_{Air,In}}}{1+(1+2\beta)\frac{N_{stack}I}{2FQ_{Air,In}}} \tag{5.51}$$

$$P_{Oxy,Ca,Out} = P_{Ca,Out}\frac{\left(\lambda_{Oxy}-1\right)\left(1-x_{W,atm}\right)}{\lambda_{Oxy}+0.21(1+4\beta)\left(1-x_{W,atm}\right)} \tag{5.52}$$

The air mass flow $Q_{Air,In}$ into the air compressor is sum of the dry air mass flow $Q_{DryAir,In}$ and water vapor mass flow $Q_{Vapor,In}$, as Eq. 5.49. The dry air mass flow is determined by output current I and cell number N_{stack} of the fuel cell stack and excess stoichiometric ratio λ_{Oxy} of air, as Eq. 5.50. According to data in Table 2.1, if the environment temperature is 35 °C and water vapor in the environment is saturated, the mole fraction of water vapor in the air is just around 5.6%. For simplicity, the mole fraction of water vapor in the air is set as 5% for modeling and analysis.

According to Eq. 3.17, the partial pressure $P_{W,Ca,Out}$ of water vapor at the cathode outlet will be increased with the increased net water flux coefficient through the membrane, the increased relative humidity of environment air and the decreased excess stoichiometric ratio of air. Substituting Eq. 5.50 into Eq. 3.17, the partial pressure of water vapor at the cathode outlet is presented as Eq. 5.51. To be noticed is that all water leaving the cathode chamber is assumed to be in the water vapor form. If water vapor at the cathode outlet is not saturated, the partial pressure $P_{Oxy,Ca,Out}$ of oxygen at the cathode outlet is presented as Eq. 5.52. The partial pressure of oxygen will be increased with the increased stoichiometric ratio of air, the decreased relative humidity of the environment air and the decreased net water flux coefficient through the membrane.

Based on Eqs. 5.51–5.52, the partial pressure of oxygen and water vapor at the cathode outlet will be increased at the same time with the increased gas pressure at the cathode outlet. It's obvious that the excess stoichiometric ratio of air is target variable of the close-loop control algorithm of the air supply subsystem. Even though the net water flux coefficient and relative humidity of environment air are not known, their contribution to partial pressure of oxygen and water vapor at the cathode outlet is limited. The net water flux coefficient is negative and the water is transported from cathode to anode.

If the fuel cell is operated at 60 °C, the net water flux coefficient β is 0 and the mole fraction of water vapor in the air is 5%, the highest saturation of water vapor and the corresponding lowest oxygen partial pressure are presented with respect to the excess stoichiometric ratio of air and gas pressure at the cathode outlet (Fig. 5.5a, c). If the fuel cell is operated at 60 °C, the net water flux coefficient β is −0.1 and the mole fraction of water vapor in the air is 0%, the saturation of water vapor and

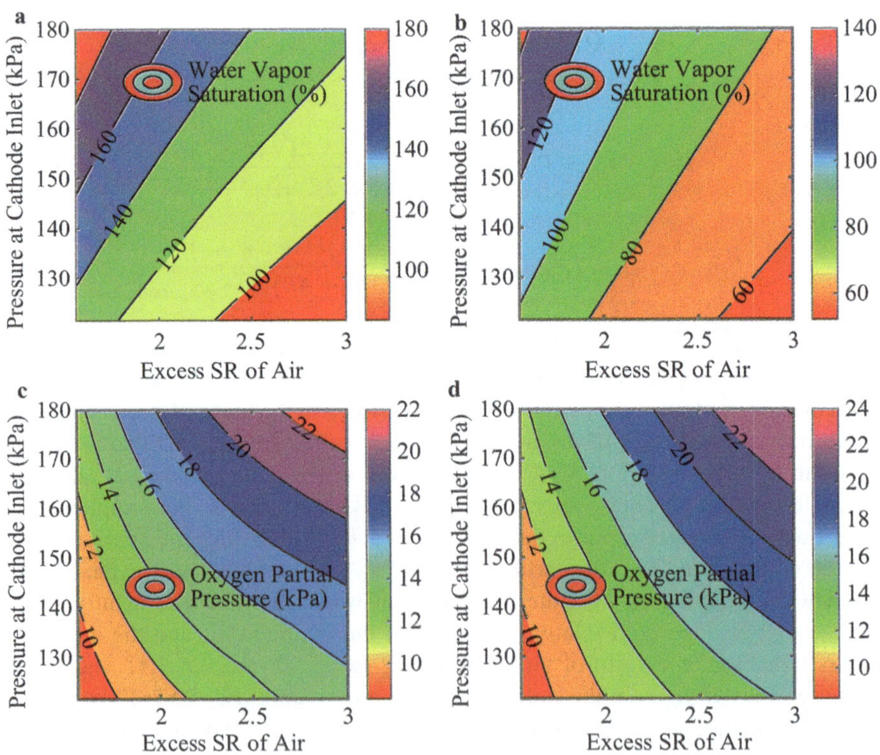

Fig. 5.5 Gas component at cathode outlet: **a** The highest saturation of water vapor, **b** The saturation of water vapor with dry supplied air, **c** The lowest partial pressure of oxygen, **d** The partial pressure of oxygen with dry supplied air

the corresponding oxygen partial pressure are presented with respect to the excess stoichiometric ratio of air and gas pressure at the cathode outlet (Fig. 5.5b, d).

Most importantly, the equipotential line of oxygen partial pressure and that of water vapor partial pressure are almost perpendicular to each other and optimization of water content and performance of the fuel cell is based on this phenomenal. Along the equipotential line of oxygen partial pressure, saturation of water vapor is decreased with increased excess stoichiometric ratio of air and it implies better capability of water removal of exhausted air at the cathode outlet.

5.4.2 Method for Searching Optimized Operating Condition

Even though estimation of partial pressure of oxygen and water vapor is effective feedback for water management strategy of the fuel cell system, they are not measurable in practice. After setting operating condition like excess stoichiometric ratio, operating temperature and gas pressure at the cathode inlet, output voltage of the fuel cell will be influenced by water content in the fuel cell. Drying-out or flooding will lead to decreased performance of the fuel cell definitely. If the high frequency resistance is not available, water content and oxygen partial pressure are implicitly related to gas pressure at the cathode outlet.

A method for searching optimized operating condition is proposed (Fig. 5.6) instead of the direct estimation, according to data and their variation trend in Fig. 5.5. At the cathode outlet, the partial pressure of oxygen satisfies the inequality as $P_3 > P_2 > P_1$ and the saturation of water vapor satisfies the inequality as $S_3 > S_2 > S_1$. Initially, the fuel cell is operated as the point A. If the net water flux coefficient keeps constant, the operating condition of the fuel cell can be moved along the equipotential line of

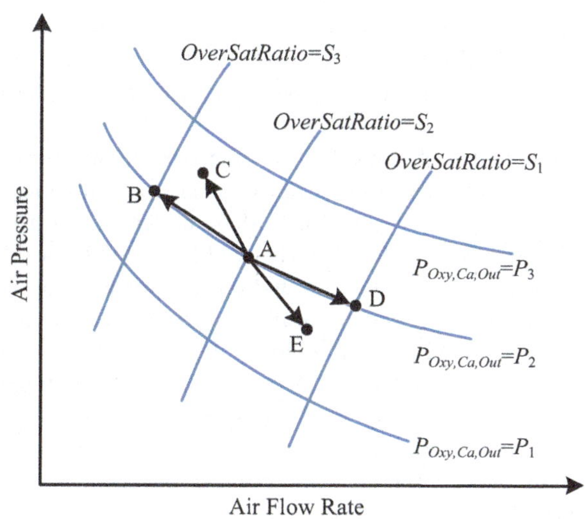

Fig. 5.6 Method for searching optimized operating condition

Table 5.1 Output voltage of the fuel cell with different operating condition

Situation	A→C	C→A	A→E	E→A
100% > S3	Increased voltage	Decreased voltage	Decreased voltage	Increased voltage
S3 > 100% ≥ S2	Similar voltage	Similar voltage	Decreased voltage	Increased voltage
S2 > 100% ≥ S1	Decreased voltage	Increased voltage	Similar voltage	Similar voltage
S1 > 100%	Decreased voltage	Increased voltage	Increased voltage	Decreased voltage

oxygen pressure like the point B and point D. If the operating condition is changed from point A to point B, gas pressure at the cathode outlet is increased but the excess stoichiometric ratio of air is decreased. If the operating condition is changed from point A to point D, gas pressure at the cathode outlet is decreased but the excess stoichiometric ratio of air is increased.

In fact, variation of operating condition leads to variation of net water flux coefficient. If the operating condition is changed from point A to point B, the final oxygen partial pressure and saturation of water vapor are moved to point C due to decreased net water flux coefficient through the membrane. If the operating condition is changed from point A to point C, the final oxygen partial pressure and saturation of water vapor are moved to point E due to increased net water flux coefficient. The corresponding voltage variation along with moving operating condition from point A to point C, back to point A, then to point E and again back to point A is summarized as Table 5.1.

For the first situation with S_3 smaller than 100%, if the operating condition is changed from point A to point C, the voltage at point C is larger than that at point A and once the voltage at point C reaches its maximal, it begins to decrease at a slower speed than that at point A. If the operating condition is changed from point C to point A, the voltage at point A is smaller than that at point C and once the voltage at point A reaches its maximal, it begins to decrease at a faster speed than that at point C. That is to say, after the procedure is executed from point A to point C and back to point A, it is obvious that the operating condition should be set on left side of the point C and there is no need to execute the following procedure.

For the second situation with S_3 larger than 100% and S_2 smaller than 100%, if the operating condition is changed from point A to point C, the voltage at point C is close to that at point A and once the voltage at point C reaches its maximal, it begins to decrease at the similar speed as that at point A. If the operating condition is changed from point C to point A, the voltage at point A is close to that at point C and once the voltage at point A reaches its maximal, it begins to decrease at the similar speed at that at point C. That is to say, after the procedure is executed from point A to point C and back to point A, it is obvious that the operating condition should be set between point A and point C and there is no need to execute the following procedure.

For the third situation with S_2 larger than 100% and S_1 smaller than 100%, if the operating condition is changed from point A to point C, the voltage at point C is smaller than that at point A and once the voltage at point C reaches its maximal, it begins to decrease at a faster speed that that at point A. If the operating condition is changed from point C to point A, the voltage at point A is larger than that at point C and once the voltage at point A reaches its maximal, it begins to decrease at a slower speed than that at point C. That is to say, after the procedure is executed from point A to point C and back to point A, it is obvious that the operating condition should be set on right side of point A. The transition of operating condition from point A to point E and back to point A follows the same trend as that in the second situation.

For the fourth situation with S_1 larger than 100%, if the operating condition is changed from point A to point C, the voltage at point C is smaller than that at point A and once the voltage at point C reaches its maximal, it begins to decrease at a faster speed that that at point A. If the operating condition is changed from point C to point A, the voltage at point A is larger than that at point C and once the voltage at point A reaches its maximal, it begins to decrease at a slower speed than that at point C. That is to say, after the procedure is executed from point A to point C and back to point A, it is obvious that the operating condition should be set on right side of point A. The transition of operating condition from point A to point E follows the same trend as that from point C to point A and it is obvious that the operating condition should be set on right side of point E.

In hence, once the performance of the fuel cell at two points is similar to each other and the performance away from the two points is worse than the former, the optimized operating condition shall lie between the two points. More experiment can be performed to further optimize the operating condition between the two point and the Bang-Bang control is effective for this searching method (Fig. 5.7). To be specific, the target output voltage of the fuel cell is not known initially and through designated procedure of changing operating condition, the optimized operating condition can be found based on variation of the output voltage, like the maximal voltage, the time when the voltage reaches the maximal, and the speed when the voltage begins to decrease. Usually, the decrease of voltage may continue for hundreds of seconds. For system application, dynamic response of the searching method cannot satisfy the requirement and the high frequency resistance can be utilized.

5.4.3 Water Management Strategy and Experiment Validation

The water management strategy for the fuel cell system is presented in detail (Fig. 5.8). Firstly, the specialized DC/DC converter proposed in Sect. 3 injects current excitation at target frequency to the fuel cell stack once receiving command from the system controller. The cell voltage monitor is triggered to calculate the AC impedance at target frequency based on the sampled voltage response and the current excitation signal. The cell voltage monitor also samples the actual dynamic voltage of each cell and the stack.

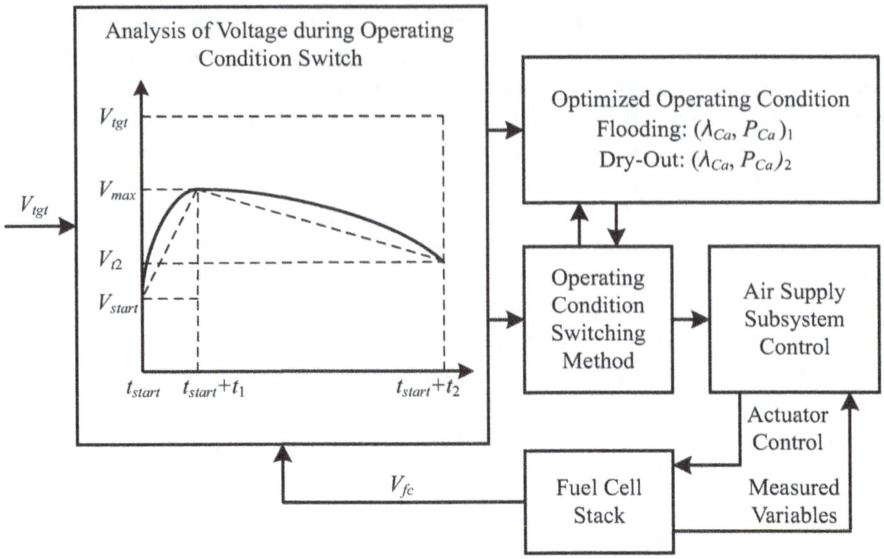

Fig. 5.7 Integrated method for searching optimized operating condition and adjustment of water content

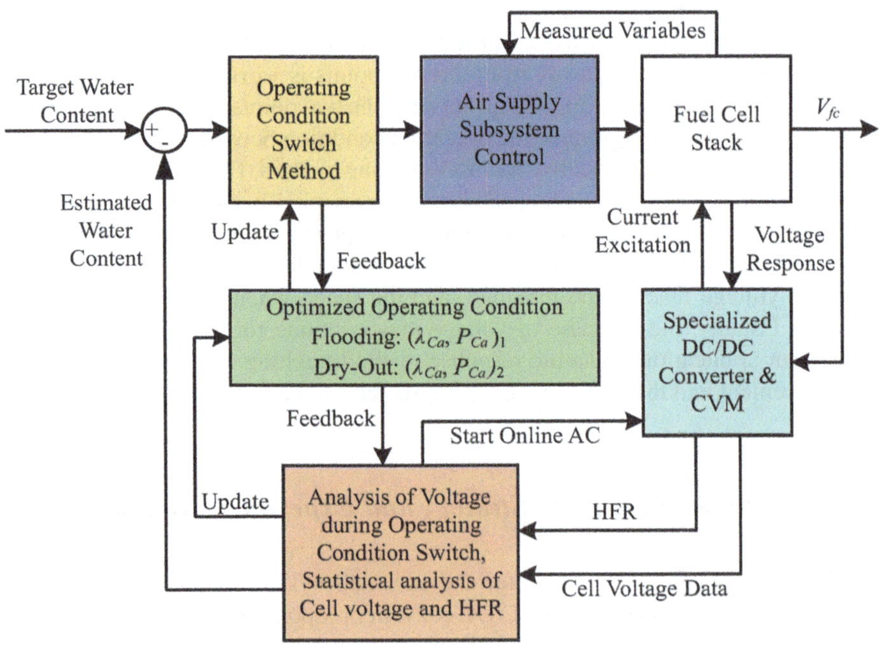

Fig. 5.8 Close-loop control of water content of a fuel cell

Secondly, according to the model of water content of MEA in Sect. 5.2 and the water content estimation method in Sect. 5.4, the dynamic response of the fuel cell voltage, statistical indexes for cell voltage and the high frequency resistance are analyzed to estimate the water content inside the fuel cell.

Thirdly, the estimated water content is compared with the target water content and based on their difference and the switching method for operating conditions, the stoichiometric ratio and operating pressure of the supplied air is changed. Fourthly, based on the dynamic model of the air supply subsystem, the control algorithm guarantees that the fuel cell stack is operated under target conditions specified in the third step.

Finally, during dynamic control of water content, the analysis of fuel cell voltage based on the switching method and the high frequency resistance contributes to quantifying the searching method for optimizing operating conditions of the fuel cell and the stack.

The experiment is performed on a fuel cell stack with 120 cells in serial and the effective surface area is 287 cm^2. The output current is 290 A, the operating temperature 65 °C, the initial operating pressure at the cathode inlet 150 kPa and the stoichiometric ratio of supplied air is 2.0. There is no humidifier in the air supply subsystem. The operating pressure at the anode inlet is controlled to follow the dynamic pressure at the cathode inlet. The stoichiometric ratio of supplied hydrogen is 1.5 and no hydrogen recirculation is considered. The fuel cell is connected to the specialized DC/DC converter and the electronic load as Fig. 3.4.

After multi-iteration of development and test, the specialized DC/DC converter is miniaturized and designed to have fast response and low power dissipation for system test. To be noticed is that only the high frequency resistance at 400 Hz is measured. The high frequency resistance can be measured every 2 s. The cell voltage is measured every 300 ms.

The experiment result is shown (Fig. 5.9), including the stoichiometric ratio and pressure at the cathode inlet of the supplied air, the average high frequency resistance of cells, the average cell voltage and the voltage difference between the highest and the lowest cell voltage. The analysis is elucidated as follows.

(1) The fuel cell is operated under the A1 condition for 20 min. The pressure at the cathode inlet is 150 kPa and the stoichiometric ratio of supplied air 2. The average high frequency resistance is decreased, the average cell voltage is increased and then decreased with oscillation, and the voltage difference is increased gradually, then decreased significantly and finally increased. It tells that the fuel cell stack is in the flooding state and water must be removed.

(2) From 66,200 s, the operating condition of the fuel cell is switched to the A2 condition. The pressure at the cathode inlet is 140 kPa and the stoichiometric ratio of supplied air 2.2. The average high frequency resistance is increased significantly and then it tends to be stable. The average cell voltage is increased and then decreased gradually. The voltage difference is increased significantly and then decreased gradually. It tells that the fuel cell stack is in the drying out state and water recovering is needed.

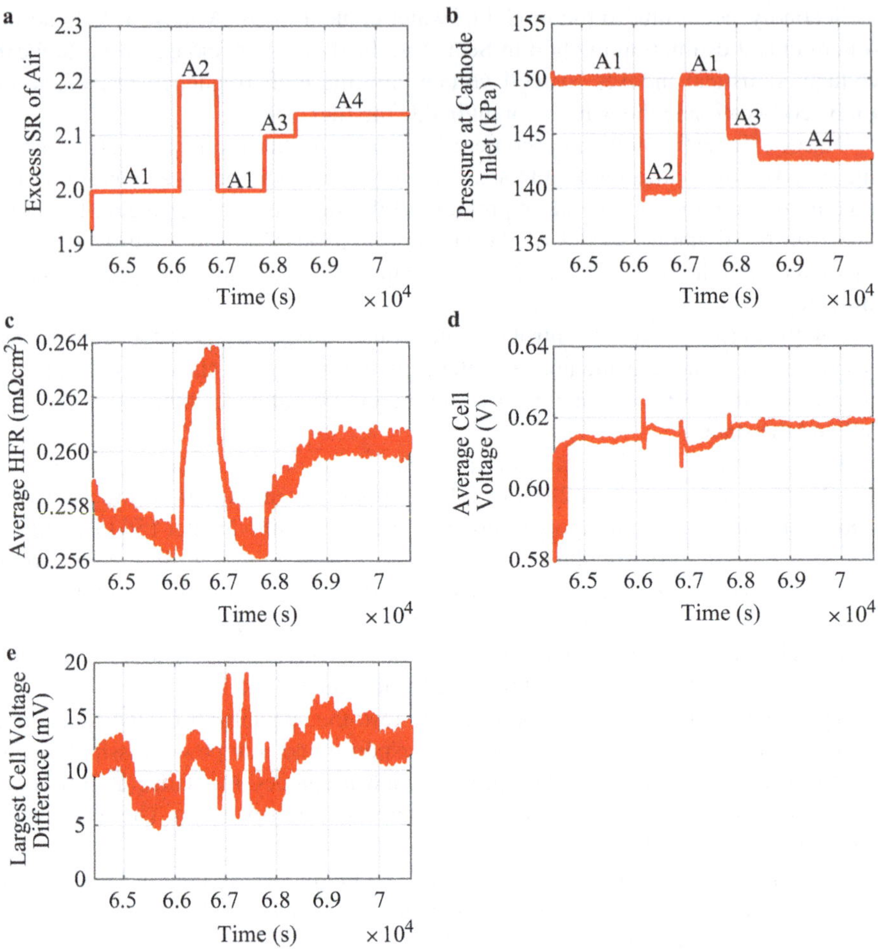

Fig. 5.9 Experiment result of water content management: **a** Excess stoichiometric ratio with time, **b** Pressure of supplied air at cathode inlet with time, **c** The average HFR of the fuel cell stack with time, **d** The average cell voltage of the fuel cell stack with time, **e** The largest cell voltage difference with time

(3) From 66,800 s, the operating condition of the fuel cell is switched to the A1 condition again. It is confirmed that the fuel cell is in the flooding state under the A1 condition. The average high frequency resistance is decreased significantly and then it tends to be stable. The average cell voltage is decreased very fast and then increased gradually after almost 1,000 s. The voltage difference oscillates significantly for the initial time and then it tends to be stable as the first test under the A1 condition.

(4) From 68,500 s, the operating condition of the fuel cell is switched to the A3 condition. The pressure at the cathode inlet is 145 kPa and the stoichiometric

ratio of supplied air 2.1. The average high frequency resistance is increased but the slope versus time is decreased gradually and then it tends to be stable. The average cell voltage is increased gradually and decreased gradually. In comparison with result under the A2 condition, the average cell voltage in this step is more stable. The voltage difference is increased and then it decreases slowly. It tells that the fuel cell stack is in the slight flooding state but it's appropriate.

(5) From 68,500 s, the operating condition of the fuel cell is switched to the A4 condition. The pressure at the cathode inlet is 142.5 kPa and the stoichiometric ratio of supplied air 2.15. The average high frequency resistance is increased during the following 500 s and then it maintains at 0.26 mΩ automatically for 30 min. The average cell voltage is slightly larger than that under the A3 condition and it is the highest voltage over these four groups of operating condition. The voltage difference is increased but it reaches a stable state. It tells that the fuel cell stack is in a good state.

References

1. Pukrushpan JT, Peng H, Stefanopoulou AG (2004) Control-oriented modeling and analysis for automotive fuel cell systems. J Dyn Syst Measur Control Transa ASME 126(1):14–25
2. Jeffrey LF (2011) Compressor performance map generation and testing per SAE J1723. California Polytechnic State University, U.S.A.

About the Author

Po Hong received his Ph.D degree from the Tsinghua University in 2019, and his B.S. degree from the Tsinghua University in 2013, all in automotive engineering. He holds 26 Chinese patents, 1 US patent and 2 PCT application, and published 12 journal and conference papers. He received the Excellent Ph.D Thesis Paper Award and SHUIMU Scholar Award in 2019.He studies at Tsinghua University as a post doctor from 2019 until now.

P. Hong, *Water Content Estimation and Control of PEM Fuel Cell Stack and the Individual Cell in Vehicle*, Springer Theses, https://doi.org/10.1007/978-981-16-8814-0

The manufacturer's authorised representative in the EU is Springer Nature Customer Service Centre GmbH, Europaplatz 3, 69115 Heidelberg, Germany. If you have any concerns regarding our products, please contact ProductSafety@springernature.com

Printed and bound by CPI Group (UK) Ltd, Croydon, CR0 4YY
30/09/2026
02225307-0011